INTRODUCTION TO INSECT PEST MANA~
Robert L. Metcalf and William H. Luckman, Editors

OUR ACOUSTIC ENVIRONMENT
Frederick A. White

ENVIRONMENTAL DATA HANDLING
George B. Heaslip

THE MEASUREMENT OF AIRBORNE PARTICLES
Richard D. Cadle

ANALYSIS OF AIR POLLUTANTS
Peter O. Warner

ENVIRONMENTAL INDICES
Herbert Inhaber

THE URBAN COSTS OF CLIMATE MODIFICATION
Terry A. Ferrar, Editor

THE URBAN COSTS OF CLIMATE MODIFICATION

EDITED BY

TERRY A. FERRAR

DIRECTOR,
ENVIRONMENTAL POLICY CENTER
ASSOCIATE PROFESSOR OF ECONOMICS,
THE PENNSYLVANIA STATE UNIVERSITY
UNIVERSITY PARK, PENNSYLVANIA

A WILEY-INTERSCIENCE PUBLICATION

JOHN WILEY & SONS
NEW YORK • LONDON • SYDNEY • TORONTO

Library of Congress Cataloging in Publication Data:

Main entry under title:

The Urban costs of climate modification.

  (Environmental science and technology)
  "A Wiley-Interscience publication."
  Bibliography: p.
  1. Weather control—Addresses, essays, lectures.
2. Urban climatology—Addresses, essays, lectures.
I. Ferrar, Terry A., 1946–

QC928.U7   301.31   76-23288
ISBN 0-471-25767-2

Printed in the United States of America

10 9 8 7 6 5 4 3 2 1

CONTRIBUTORS

Ralph C. d'Arge
Department of Economics
University of Wyoming
Laramie, Wyoming

Thomas D. Crocker
Department of Environmental
  Economics
University of Wyoming
Laramie, Wyoming

Joseph L. Goldman
Technical Director
International Center for Solution of
  Environmental Problems
Houston, Texas

J. Eugene Haas
Department of Sociology
Institute of Behavioral Science
University of Colorado
Boulder, Colorado

Irving Hoch
Resources for the Future
Washington, D.C.

Jon P. Nelson
Department of Economics
The Pennsylvania State University
University Park, Pennsylvania

John H. Niedercorn
Department of Economics
University of Southern California
Los Angeles, California

Hans Panofsky
Evan Pugh Research Professor
  of Atmospheric Sciences
The Pennsylvania State University
University Park, Pennsylvania

Clarence M. Sakamoto
Center for Climatic and Environ-
  mental Assessment
Federal Building
Columbia, Missouri

Peter G. Sassone
Department of Economics
Georgia Institute of Technology
Atlanta, Georgia

Richard J. Tobin
Department of Political Science
State University of New York at
  Buffalo
Buffalo, New York

FOREWORD

Perhaps more than any other natural phenomenon, climate has defined and limited man's habitat. As Edholm and others have recognized, without human adaptation and invention man would be constrained to live in a very small latitudinal zone of the planet.[1] Through innovation man has spread to every niche of the globe. How man and his tools, including domesticated crops, have adjusted to climate and how they might adapt more efficiently has been a matter of both speculation and scientific concern for many centuries.[2] The past few decades have brought first a realization that around large cities ambient temperatures and other climatic indicators differ from those of the surrounding countrysides.[3] Almost simultaneously, based on scientific advances and ad hoc efforts, it was hypothesized that man might be able to significantly influence climatic patterns, particularly seasonal rainfall in particular locales. In recent years the regional occurrence and spreading of air pollution, as well as recognition of the interdependence between ground cover and microclimates, have encouraged speculation that man—by conscious choice or inadvertently through economic activities—might be subtly altering what he had viewed as an unalterable decree of God or nature, that is, the climate in which he resides. Most recently, the growing complexity of our technology, particularly in the field of synthetic chemicals, an increase in the observed accumulations of poisonous elements in the environment, and the development of the global society's exotic electronics have led to a growing concern as to whether these nonnatural introductions may cause environmental, and thereby economic and social, repercussions. Johnston in 1971 hypothesized that large fleets of SSTs or subsonic aircraft flying in the stratosphere may induce reductions in the ozone profile through chemical interaction between ozone and nitrogen oxides emitted by aircraft in the stratosphere, thereby allowing an increase in the amount of ultraviolet light reaching the earth's surface.[4] Ultraviolet light induces sunburn (and suntan) to those exposed; and some preliminary scientific evidence suggests that it may, through cumulative processes, increase the probability of both fatal and nonfatal forms of skin cancer.[5] In addition to the potential effects on people, the useful lifetime of certain types of chemical materials may be adversely affected by enhanced ultraviolet light.[6]

Stratospheric flight by relatively large fleets of planes may also reduce the average global temperature because of a shielding effect caused by the concentrated emissions of sulfur oxides and particulates at altitudes (12 to 20 km) so high that they require a great deal of time to return to the earth's surface.[7] While such an impact is barely beyond scientific conjecture, the impacts of a long-term cooling trend could be potentially disastrous.[8] The global economic costs of a 1°C reduction in mean temperature might well exceed the value of the current gross national product of the United States.[9]

Just recently, a new concern has arisen as to the ultraviolet and temperature effects of chlorofluoromethanes (commonly called fluorocarbons), which are used in a variety of consumer products ranging from refrigerators to hairsprays. Some scientists have proposed that these relatively inert compounds ultimately accumulate in the stratosphere and then interact photochemically to alter the ozone profile within the stratosphere, thus altering the amount of ultraviolet radiation contacting the earth's surface and also affecting global temperatures.[10] A further conjecture is that the nitrogen fertilizers used on farms may reach the stratosphere with similar photochemical impacts and resulting surface changes.[11] None of these scientific observations has been "proven," but there appears to be enough evidence—laboratory, theoretical, and other—to concern society. Perhaps the stratosphere is the first of a whole sequence of higher atmospheric regions that can be modified by man's production and consumption decisions on earth. We must now contemplate a basic interdependence between man and his environment on more than a local or regional scale.

This book brings together chapters that constitute a first step toward understanding the economic and social reactions and effects of an altered urban climate. The studies are not definitive but they do indicate how man should value decisions on altering his urban climatic regimen. They collectively also provide the first advancement beyond a long series of rather loose observations on the economic value of climate to urban communities—ranging from heating to snow removal costs—and how these are affected by climatic variations. These methodologies and empirical estimates will provide at least partial insight into the benefits and costs of humanly induced climate change. Thus these studies will be helpful, and probably definitive (for a while), in the debate during the next decade on regulating pollutant emissions affecting climate, whatever their source, on both a national and global basis. This book provides a first step toward evaluating the feedback effect of climate on urban man within the man-on-climate and climate-on-man chain of effects. However, only with continued dedicated research will both of these links be understood sufficiently to become highly valuable for public policy decisions.

Terry Ferrar has brought together a truly interesting and relatively complete set of original research papers on the relationship of the urban economy to its ambient climate. It is almost a French menu of potential effects that most scientists (with our vision of today) would consider inclusive of the major impacted sectors. He has also collated an interesting set of descriptions of man's climatic history dating from before the birth of recordkeeping. This collection of papers should become the benchmark for future economic and social valuations and, in fact, they have already become so in evaluating the Concorde and fluorocarbon production and use in the United States.

RALPH C. D'ARGE
JOHN S. BUGAS,
Professor of Economics,
Chairman, Panel on Economic and Social Impacts,
Climatic Impact Assessment Program
Department of Transportation

July, 1976

REFERENCES

1. O. G. Edholm, "Problems of Acclimatization in Man," *Weather* **21**, 340–350 (1966).

2. H. E. Landsberg, "The Assessment of Human Bioclimate: A Limited Review of Physical Parameters," Technical Note 123, World Meteorological Organization, Geneva, Switzerland (1972).

3. W. P. Lowry, "Climate of Cities," *Scientific American* **217**, 15–23 (August 1967).

4. H. S. Johnston, "Reduction of Stratospheric Ozone by Nitrogen Oxide Catalysts from Supersonic Transport Exhaust," *Science* **173**, 517–522 (1971).

5. F. Urbach, "Field Measurements of Biologically Effective Ultraviolet Radiation and Its Relation to Skin Cancer in Man," Proceedings of the Third Conference on the Climatic Impact Assessment Program, Report No. DOT-TSC-OST-74-15, U.S. Department of Transportation (1974).

6. A. R. Schultz, D. A. Gordon, and W. L. Hawkins, "Materials Weathering," CIAP Monograph 6, U.S. Department of Transportation, Chap. 3.10, pp. 239–248 (September 1975).

7. J. A. Coakley, Jr., and S. H. Schneider, "Possible Climatic Effects of Supersonic Transports," National Center for Atmospheric Research, Boulder, Colorado (1974).

   B. M. Herman, S. R. Browning, and R. J. Curran, "The Effect of Atmospheric Aerosols on Scattered Sunlight," *Journal of Atmospheric Science* **28**, 419 (1971).

   F. M. Luther, "Solar and Longwave Effects of Stratospheric Perturbations," Report UCRL-76106, Lawrence Livermore Laboratory, California (Oct. 1, 1974).

   V. Ramanathan, "A Simplified Stratospheric Radiative Transfer Model: Theoretical Estimates of the Thermal Structure of the Basic and Perturbed Stratosphere," Paper presented at the AIAA Second International Conference on the Environmental Impact of Aerospace Operations in the High Atmosphere, San Diego, California (July 8–10, 1974).

8. R. d'Arge (ed.), *Economic and Social Measures of Biologic and Climatic Change,* CIAP Monograph 6, Chap. 1, Report No. DOT-TST-75-56, U.S. Department of Transportation (Sept. 1975).

9. R. d'Arge, "Economic Analyses of Pollution Resulting from Stratospheric Flight: A Summing Up," prepared for the Fourth Climatic Impact Assessment Program, DOT, Scientific Report Series (April 1976).

10. M. J. Molina and F. S. Rowland, "Stratospheric Sink for Chlorofluoromethanes: Chlorine Atom Catalyzed Destruction of Ozone," *Nature* **249,** 810 (1974).

    P. J. Crutzen and S. A. Isaksen, "The Impact of the Chlorocarbon Industry on the Ozone Layer," paper submitted to the *Journal of Geophysical Research*.

11. Michael McElroy, testimony before the Federal Intraagency Task Force on Inadvertent Modification of the Stratosphere (Feb. 1975).

SERIES PREFACE

Environmental Sciences and Technology

The Environmental Sciences and Technology Series of Monographs, Textbooks, and Advances is devoted to the study of the quality of the environment and to the technology of its conservation. Environmental science therefore relates to the chemical, physical, and biological changes in the environment through contamination or modification, to the physical nature and biological behavior of air, water, soil, food, and waste as they are affected by man's agricultural, industrial, and social activities, and to the application of science and technology to the control and improvement of environmental quality.

The deterioration of environmental quality, which began when man first collected into villages and utilized fire, has existed as a serious problem since the industrial revolution. In the last half of the twentieth century, under the ever-increasing impacts of exponentially increasing population and of industrializing society, environmental contamination of air, water, soil, and food has become a threat to the continued existence of many plant and animal communities of the ecosystem and may ultimately threaten the very survival of the human race.

It seems clear that if we are to preserve for future generations some semblance of the biological order of the world of the past and hope to improve on the deteriorating standards of urban public health, environmental science and technology must quickly come to play a dominant role in designing our social and industrial structure for tomorrow. Scientifically rigorous criteria of environmental quality must be developed. Based in part on these criteria, realistic standards must be established and our technological progress must be tailored to meet them. It is obvious that civilization will continue to require increasing amounts of fuel, transportation, industrial chemicals, fertilizers, pesticides, and countless other products and that it will continue to produce waste products of all descriptions. What is urgently needed is a total systems approach to modern civilization through which the pooled talents of scientists and engineers, in cooperation with social scientists and the medical profession, can be focused on the development of order and equilibrium to the presently disparate segments of the human environment. Most of the skills and tools that are needed are already in existence. Surely a technology that has created such manifold environmental problems is also capable of solving them. It is our hope that this Series in Environmental Sciences and Technology will not only serve to make this challenge more explicit to the established professional but that it also will help to stimulate the student toward the career opportunities in this vital area.

Robert L. Metcalf
James N. Pitts, Jr.
Werner Stumm

PREFACE

The human race has always recognized its dependence on the atmospheric environment. Whether we discuss the caveman society of prehistoric times or the sophisticated space-suit technology of the twentieth century, this lifeline linkage has always influenced behavior. It is clear that patterns of human settlement; clothing design; and agricultural, recreational, and even cultural activities reflect man's dependence on the environment.

Modern man has found it possible to mobilize various combinations of resources to use the atmosphere to enhance his well-being. The discipline of meteorology has recently demonstrated, however, that this linkage is bilateral. Hence man's behavior is not only influenced by the atmosphere, but his activities now affect the character of the atmosphere itself.

As modern technological production and distribution processes continue to advance and as the world's population continues to concentrate in various regions of the available land masses, environmentalists and climatologists warn that significant modifications in climatic conditions are likely to result. For example, it is suggested by some that man's injection of particulate and aerosol matter into the atmosphere will produce a cooling of our planet because of the increased reflection of the sun's rays. Others, however, argue that the increasing quantity of atmospheric carbon dioxide is a more significant problem. The carbon dioxide molecule has an affinity for ozone, a very important ultraviolet shielding material in the stratosphere. Increasing the concentration of atmospheric carbon dioxide is expected to reduce this ultraviolet shielding, thus resulting in an increase in the ultraviolet flux falling on the earth's crust. The opposing nature of these theories, along with the ultimate impact potential on our way of life, is illustrative of the gravity of the issues addressed in this book.

This collection of original works examines the economic consequences of such climatic modifications with respect to the important and somewhat unique problems of the urban sector. The first part of this effort presents an overview and characterization of the magnitude of the problem. In Chapter 1 Clarence Sakamoto reviews the history of man-induced and natural climatic change and its influence on man's early attempts to urbanize. Hans Panofsky in Chapter 2 surveys the variety of influences contemporary man imposes on the atmosphere and, using examples, argues that the rate of climatic change is increasing in both intensity and scale. In Chapter 3 Joseph Goldman demonstrates how increased convection caused by flow obstructions from tall buildings and heat radiation from concrete, together with enhanced seeding by aerosol emissions, is changing the pattern of rainfall in urban areas. Chapter 4, by Jerome Haas, reviews the many social implications resulting from climatic changes.

The next studies were designed to provide both a methodological framework and a quantitative evaluation of the effects associated with parametrically defined climatic perturbations. These studies present the reader with an illustration of state-of-the-art techniques for empirically evaluating the economic consequences of short- to medium-run climatic modifications. These works are not concerned with long-run influences for two reasons. First, the problems of forecasting what the climate will be like in the distant future are quite large. The usual result of such speculation is the conclusion of ultimate catastrophe, derived by projecting exponential growth curves. Therefore the lack of precision in climatic parameter ranges has discouraged us from this exercise. Second, a person who speaks of the long-run climatological time frame must recognize the likelihood of fundamental institutional change. To analyze the economics of climatic modification, we must restrict ourselves to cost measures relevant to the given institution setting.

In Chapter 5 Jon Nelson explores, through the use of an econometric model, the impact that climatic perturbations can have on the demand for fossil fuels. Thomas Crocker then demonstrates in Chapter 6 the importance of climatic influences on electricity consumption and the associated money costs of climatic modifications. In the following chapter Crocker illustrates how economic theory can be used to infer the values consumers attach to alternative climates. In Chapter 8 Irving Hoch, using regression analysis, relates the money wage of a locality to its climatic variables and estimates the effect of climatic changes on real income and general welfare.

In addition to the private sector impacts resulting from climatic changes, the next two chapters explore the implications of such changes on the public sector. In Chapter 9 Peter Sassone examines public municipal expenditures that are sensitive to climatic perturbations and attempts to assess the magnitude of costs associated with such changes. Richard Tobin then focuses on the political problems and consequences local governments are likely to confront as a result of climatic changes.

Finally, John Niedercorn pursues the broader regional approach to the economic consequences of climatic changes. Specifically, by estimating the net capital costs that might be incurred by a southward shift in the U.S. Corn Belt, the author examines whether a worldwide problem resulting from climatic changes exists.

The preface presents the editor with the opportunity to formally express his sincere appreciation to the many persons who were instrumental in the preparation process. The reader will immediately appreciate that the contributing authors invested considerable time and personal resources during the preparation of their manuscripts, and the fruits of their efforts are eloquently presented in this volume.

Many of the papers presented herein, and indeed the germination of this book, were derived from a major research investigation sponsored by the U.S. Department of Transportation. This study, commonly known as the Climatic Impact Assessment Program, was under the general direction of Alvin Grobecker. I had the fortunate experience of being drafted to assist in the economic portion of this study by Dr. Ralph d'Arge, and it was through this project that I became acquainted with many of the authors.

In terms of the administration and editorial support during the final preparation efforts, it is with great pleasure that I recognize the role of my university staff, particularly Alan B. Brownstein and James Ford.

It has been professionally stimulating and enjoyable to associate with the scholars who contributed to this edition. My hope is that the collective presentation of these original manuscripts will contribute to advancing society's understanding of these important and timely issues.

TERRY A. FERRAR

University Park, Pennsylvania
July 1976

CONTENTS

1

Historical Survey
of Climatic Change and Its
Relationship to Man

CLARENCE M. SAKAMOTO
Center for Climatic and
Environmental Assessment
Federal Building
Columbia, Missouri

The possibility of man-induced or natural climatic change has posed many questions concerning the ability of man to cope with the changing environment. Responses to these changes are difficult to predict or assess because of the complexity of interacting sociological, economic, political, technological, and cultural factors. In addition, the problem is further complicated because climatic change has taken place over a span of tens to hundreds of years. One can, however, study past climatic changes to seek evidence or accounts of their effects in order to better understand the possible responses of man to future climatic changes. This chapter, therefore, discusses early historical climatic changes and the impacts they have had on man in his early urban communities.

One problem in attempting to associate historical climatic changes with the response of man to climatic change is the difficulty in proving the association between them. Interdisciplinary piecing together of evidence to form a portion of what might be considered a gigantic and complex puzzle is still being pursued by scientists today. In some cases associations can only be suggested with evidence favoring or disfavoring a hypothesis still in balance.

1

Data on climatic changes during later years are more readily available. Therefore, the changes discussed in this chapter are confined to the period since about 5000 B.C., the period in climatic history known as the *optimum,* a relatively warm period. Cooling since that period has been evident.

Naturally occurring climatic changes have had a major impact on past communities. These early civilizations were agriculturally oriented; hence, their economic success or failure depended much on the tide of the climate. Year to year fluctuations in the weather could be coped with, but long term changes that covered tens to hundreds of years slowly eroded the capacity of a community to produce a successful crop. If man was unable to meet this change, famine, economic chaos, armed conflicts, and as history relates to us, complete abandonment of communities resulted.

History has revealed the consequences of repeated crop failures. This should lead one to ponder the food crises of the mid-1970s. Already famine has struck in many areas of the world; it has brought death to thousands; it has initiated economic instability and uneasiness among people in the world.

Evidence for Climatic Change

How, one may ask, is it possible to estimate past climatic changes quantitatively? Fortunately, several methods are available or have been developed and utilized by scientists and historians. These range from the study of historical documents to excavation of remains of old ruins and applying carbon 14 or pollen analysis techniques.

Uncovered artifacts have provided evidence that related a particular type of handicraft to a particular period of history during which the handicraft was prominent. Building materials, such as bricks, have also provided an indication of the prevailing climate at a particular time. Sir Mortimer Wheeler[48] uncovered literally millions of baked bricks rather than mud bricks in the old remains of Mohenjo dara and Harappa of the Indus civilization which dated about 2500 to 1700 B.C. The finding of large numbers of baked bricks is considered strong evidence of climatic change as wood is needed to bake bricks. This suggested a source from a nearby forest, indicative of a wet climate. Today, the Indus Valley bordering Pakistan and India is an arid area.

Historical accounts have also been very informative. One of the oldest and most well known is Herodotus' *History* as cited by Brooks.[6] This book provides notes indicative of the climate during the first millennium B.C. From Herodotus' description of the Caspian Sea, for example, the Aral Sea in Russia is thought to have once been connected to the Caspian Sea, indicating a much wetter climate than currently exists in that area. A summary of Chinese farming as documented in the *Chin Min Yao* and written more than 1400 years ago provides abundant phytological notes.[41] This document reveals that peach trees north of the Yellow River bloomed 15 to 30 days later than today indicating a climate much cooler than now.

Other kinds of records were kept in medieval times. Church diaries and accounting records offer additional support of the trend of climatic change during and since that time. A series of entries of the account rolls of the Bishopric of Winchester for the period 1209 to 1350 A.D. was published in 1960.[46] These records provide information of weather and wheat harvest during that period.

Another unique mode of revealing climatic change has been offered by the archeologist Henri Lhote.[29] His discovery of the Tassili rock paintings in the Sahara provided additional support that the desert was once teeming with plant, animal, and human life. Several hundreds of these paintings have provided a pictorial log of the inferred cultural and climatic changes and finally, the demise of a civilization.

Glaciologists have also contributed much information that helps to partially solve the mammouth puzzle of climatic change. In his book, Ladurie[26] compared historical notes, engravings, and paintings that provide information on the movements of famous glaciers in Europe. In 1969 Dansgaard, Johnsen, Miller, and Langway[11] analyzed the ice sheet at Camp Century, Greenland and provided evidence that confirmed the cooling that took place in Greenland about the time of the decline of the Norse colony on that island.

Pollen analysis by botanists offers another method to shed light on climatic change. The basic assumption of this method is that the climate in past years was similar to that of an area where similar plant species are found today. In addition to pollen analysis, fossilized remains of plants and animals, including bones, have also provided evidence of the kind of climate in the past.

Dendroclimatology has offered a relatively unique approach to the study of climatic change. The work of Douglass,[13] Schulman,[38] and more recently by Fritts[15] exemplifies the possibilities in this field. Interpretation of growth, however, is difficult because of the interaction of a complexity of factors, among which are temperature and precipitation. Since a tree adds one ring of growth a year, yearly information can be obtained from this procedure. Tree-ring analysis has confirmed the dry period in the southwestern United States during the late 13th century.

Other disciplines including hydrology and irrigation engineering have provided inputs into this vast study. Taylor[44] investigated the Indus Plain from an engineer's point of view by analyzing the irrigation systems and underground water. He suggested that the decline of this civilization was initiated by man-made and natural disasters, including floods while the *coup de grace* was performed by the invading Aryans about 1500 B.C.

A method that has created renewed interest in recent years is the phenological basis of studying climatic change. This method has been discussed by several investigations of climatic change including Lamb,[28] Utterstrom,[47] Ladurie,[26] and Arakawa.[2]

Phenology is the science that treats biological phenomena in relationship to climate. Ladurie[26] used phenology to relate French wine harvest dates for the years 1801 through 1885 to the mean temperatures for the period April through September. He found an excellent relationship between the two. Arakawa[2] noted in his analysis of cherry blossoms in Japan from the 9th through the 19th century that blossoming was significantly delayed for the 11th through the 14th centuries. This period, as is noted later, was a cold period in climatic history over most of the midlatitudes of the world. Chu[9] also capitalized on extensive phenological notes contained in literary documents of early China to estimate China's climate during the past 5000 years. Utterstrom[47] explains phenology further:

> The effects in fact vary greatly between different regions and different kinds of plants, according to whether temperature or rainfall is the decisive factor for the crop. . . . Phenology has taught us the period of ripening is considerably shortened within continental areas, that warmth accelerates phases of flowering and fruition, that sharp changes in temperature cause the buds to open later. . . .

It can also be said that cooler temperatures may also delay blossoming, and in some cases sharp drops in temperature may damage the fruit trees.

Climatic Change Since Optimum

Climatic Optimum

The *climatic optimum* serves as a reference in climatological history and is a generally accepted period of regional thermal maximum in modern geological times. Manley[34] used the term *hypsithermal* to characterize this period. Others, including Woodbury,[49] have referred to this period as the *alithermal*. During this epoch temperatures were as much as 2 to 3°C warmer than today. In Europe the climatic optimum is estimated roughly from 5000 to 3000 B.C. (Figure 1). In Australia the warm and humid optimum has been estimated near 5000 B.C.[16] In some parts of the world the optimum occurred earlier; in other parts it is dated later. For example, in the Bering Straits this date is approximately 8000 B.C. to 6000 B.C. In Denmark it has been estimated to occur from 2000 B.C. to 1000 B.C.[34]

The optimum was also a wetter period than today. The moist optimum in North Africa during the period 5000 to 2400 B.C. has been referred to as the *subpluvial*.[28, 8] Pollen analysis, archeological artifacts, carbon 14 dating, paintings, and historical notes have been found to suggest that the arid Sahara,[8] the

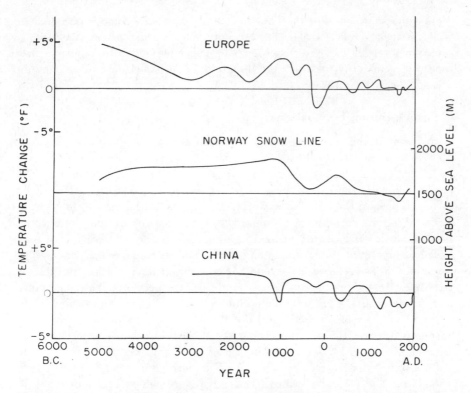

Figure 1. Temperature trend since the climatic optimum in Europe (after Brooks),[6] China (after Chu[9]) and the snow line height in Norway (after Schwarzbach[39]).

Indus Valley in northern India,[6] Italy,[41] and tropical Hawaii[22] were once moister than today. In China archeological evidence, including bones of tropical animals, has been found to suggest that 5000 to 6000 years ago the Yellow River Basin in the north central plain was as warm and wet as it is today in the Yangtze River Valley, located further south.[9]

Figure 1 shows the response of the thermal regime with time. The curves for Europe and China are the temperature changes since the climatic optimum. The smoothness of the curves reflects the lack of more frequent data, particularly during earlier times. The generalized temperature curve for Europe was drawn from historical accounts.[6] Studies by Lamb,[27] Manley,[32] and Chu[9] have supported this general curve, although fluctuations for a specific area of study have altered the curve slightly. The China curve is based on phytological studies by Chu.[9]

The snow line curve for Norway has been determined by glaciologists[39] and is based on the assumption that with warming the snow line rises, while with cooling the snow line lowers. The snow line during the climatic optimum in Norway was about 300 m above the level of today. A trend to lower elevation is evident since 1000 B.C. and corresponds to colder periods in Europe since that time. It is generally agreed that the climate near the end of the first millennium was similar to what it is today.

Post Optimum Fluctuations

In Figure 1 the curve for Europe shows a gradual temperature decline to 3000 B.C. following the climatic optimum. Temperatures rose after 3000 B.C., but began to fall sharply toward the end of the second millennium B.C. The period following the end of the climatic optimum was also characterized by increased dryness throughout the world. *Desiccation* set in. This term refers to a time of relative dryness when compared to the climate of previous periods.

It is convenient to divide changes in climate with selected periods that may have had significant effects on human cultural and economic behavior as well as on animal and vegetative life. Lamb[28] lists a chronicle of at least five periods since the climatic optimum. These include (*a*) the period of the temperature and, in some parts, a precipitation decline following the thermal maximum; (*b*) a period of a secondary optimum, the "little climatic optimum," which occurred at different times in different parts of the world but is generally dated between 400 and 1200 A.D.; (*c*) a period of cooling during the period 1200 to 1400 A.D.; (*d*) partial recovery, a warm period, lasting from about 1400 A.D. to 1500 A.D.; and (*e*) the so-called "Little Ice Age" which began in the 16th century in Europe and lasted to about the second half of the 19th century. These periods are in the most recent geological history and constitute the time for which climatic studies have been documented. Examples of these climatic periods in relationship to man are discussed later.

Climatic Decline.   Starting about 1500 B.C. (Figure 1) colder temperatures began to lead to glacier advance and freezing of previously unfrozen rivers. The cool period, generally drier, although in parts of the world considered moist, lasted nearly through the first century B.C. in parts of the world, although Oliver[36] shows this period as ending about 450 B.C. in some areas of the world. In China this period was relatively short, beginning near the end of the second millennium B.C. and lasting only for about a century.[9] In some parts of the world the cool decline came gradually; in others it came abruptly.

The abrupt decline of temperature in Europe, near 500 B.C. (Figure 1) was catastrophic to human civilizations. Around the Mediterranean cool, rainy periods filled lakes and formed bogs; alpine glaciers advanced. Storms became more numerous in the North Sea during this period and about 120 B.C. influenced the movement of the early Celts and Teutons from the western part of the German plain.[28]

Little Climatic Optimum. Climatologists have referred to the period from about 400 to 1200 A.D. as the *little climatic optimum.* This was a warm, dry period when, according to Lamb,[28] storms in the Atlantic and in the North Sea were less frequent than during the climatic decline. It was a period of exploration and settlements exemplified by the Vikings settlements in Iceland and Greenland. In comparison, the neolithic climate optimum endured for a few thousand years, while the "little" optimum lasted for a few centuries.

Little Ice Age. Another period in climatic history is frequently referred to as the *little ice age.* This was a period of pronounced deterioration around the middle of the 16th century that continued to about the middle of the 19th century.[6, 28, 32] Several minor fluctuations, some considerably colder, have occurred prior to and during this period. Like other stages of climatic history, this period was not coincident around the world. From his analysis of glaciers Ladurie[26] estimates that from 1590 to 1850 the mean temperature in Europe was probably between 0.3 to 1°C lower than the contemporary level of the 1960s.

This brief discussion of the general thermal regime sets the stage for the sections to follow. The intent is to study the possible influences of these climatic changes on the urban populations in history.

Past Urban Civilizations

Mesopotamia

The development of Mesopotamia could be considered an example of man's early attempt toward controlling the climate. New inventions and irrigation systems led this civilization in the Tigris-Euphrates valley (now Iraq), to develop the desert into a well-organized community. Ironically, it was this development of controlling the flood waters that led to the downfall of this urban community.

Sometime before 5000 B.C. there developed a culture called Ubaidans in the Iranian highlands, a region north of the Tigris River. The Ubaidans settled in the Mesopotamia flood plains but found it difficult to raise a crop because of frequent spring and summer floods. The Ubaidans, however, engineered a system of dams, reservoirs, and irrigation canals to store the excess water during heavy rains and to use it later during dry periods. With this system, the muddy flood plain was transformed into a productive farming area. Sometime during the fifth millennium B.C. a federated system of city-states followed, and the community of Sumer was found. The Sumerians practiced intense agriculture, raising cereals, animals, and fish. Several cities flourished, at times with as many as 50,000 inhabitants. New inventions, medicine, mathematics, and other fields developed, leading to a well-balanced economy. Today this same area is a sparsely populated desert; but during the climatic optimum, Mesopotamia did not resemble a desert.

What caused the demise of the Sumerian civilization? Historians claim repeated invasions by nomadic herdsmen. Huntington,[22] however, suggests that drier climate was a factor for these invasions. Huntington also believed that Mesopotamia was a desert in ancient times but later application of new innovative inventions and technical knowledge produced a vast civilization. The Sumerians also encountered repeated floods, and evidence has been found to indicate that the biblical Noah's flood took place in this area.[42] Archeological study of the soil profile shows that a great deluge, perhaps one of many prior to 3500 B.C., probably inundated Mesopotamia. Still others speculate that the salinization of the fertile soil from repeated irrigations rendered the land unproductive. Using the Salt River Valley in Arizona as an example, Taylor and Ashcroft[45] point out some major problems that can occur with sustained agriculture, based on irrigation. Twentieth century knowledge has shown that with increased salinity of water the accumulation of salts in the soil can occur. The permeability of water into the soil can be reduced by repeated irrigation as a consequence of silt and salt buildup. In addition, the structure of the soil can be altered. This in turn leads to problems related to aeration, nutrient uptake, root growth, and so forth. The demise of Mesopotamia was probably the result of a combination of factors, including floods, changing climate, and deteriorating soils.

Urbanization did not begin until man pursued his quest for food by efficient agriculture.[40] In those areas where urban communities began and expanded, adequate moisture was available through rainfall or by use of irrigation systems. But continued agricultural operations on the flood plains gradually dwindled the capacity of the once fertile soils to produce.

Susa and Jarmo

In southwestern Iran another settlement of 7000 years ago, Susa (Figure 2), was uncovered in archeological surveys.[1] Evidence indicates a permanent settlement based on dry farming prior to about 3500 B.C. This settlement roughly parallels the 300 mm precipitation isohyetal today. From carbon-14 analysis Adams[1] found in this ancient village evidence of vigorous growth that dated from about 6000 to 5500 B.C. He suggested that favorable precipitation was associated with the dense population during that period. Since irrigation systems were not known until much later, Adams suggests that the agriculture of Susa was heavily dependent on rainfall. Several centuries following 5000 B.C. rainfall was higher, but prior to 3000 B.C. drought set in and continued as late as 2000 B.C. Archeological findings suggest that the settlement declined during this dry period. Resettlement was hampered by armed conflicts and abetted by droughts, in addition to the decline of agricultural soils.

A most important contribution to world food production is rooted in the early history of the Middle East. The oldest known wheat was found by an archeological expedition at Jarmo[17] in Iraq-Kurdistan (Figure 2). The wild Emmer, *Triticum dicoccoides,* was excavated in the 1950s and dated about 6700 B.C. Helbaek[17] believes that Emmer wheat represents the earliest appearance of wheat in a cultural context. The Emmer wheat later became the cultivated wheat and spread with human migration into Asia, Africa, and Europe and maintained its lead position as a staple food crop. The Emmer wheat reached the Indus Valley during the third millennium B.C.[17] Butzer[8] believes that food producing cultures migrated rapidly when faced with prolonged drought and that drought played a part in the expansion of neolithic

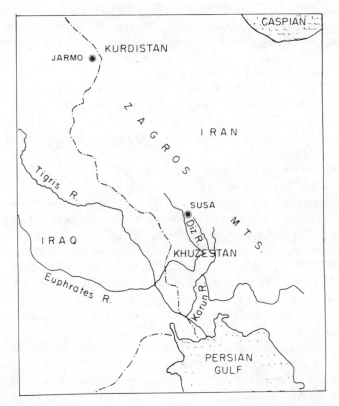

Figure 2. Tigris-Euphrates Valley, Jarmo, and Susa—early neolithic civilizations.

culture to the moister land of Europe. With human migration the introduction of a particular variety of grain into an unnatural environment led to a process of natural selection. By a series of introductions into harsher climates and with natural selection, the more advanced wheat varieties were cultivated. Although this process of selection has proceeded for centuries, there is at some point a natural environmental limit that cannot support further genetic selection. With this the species can no longer be grown.

Indus

Another civilization based on an agrarian culture flourished in what is now a desolate parched desert of the Indus Valley 4500 years ago at about the time of the climatic optimum. The heartland of this civilization was the plains region extending about a thousand miles from the Arabian Sea to the foothills of the Himalayas. The Indus River was occasionally flooded by the spring snowmelt of the Himalayas. Today, however, for most of the year only pebbles are visible in the river bottom with hardly a trickle of water. Evidence has been found to indicate that during the time of the climatic optimum rainfall was heavy.

In 1921 an archeological expedition by Sir Mortimer Wheeler[48] in Harappa, Pakistan (then northwestern India) uncovered a city that was older than any city unearthed in that area. A few years later, some 350 miles away from Harappa, a second city, Mohenjo dara, was discovered. It was estimated at one time that 40,000 people lived in each of these two cities. Now this sandy desert is considered one of the least habitable areas of the world. From about 2500 B.C., and for a thousand years, however, food raising occupied the principal economy of the cities. Unearthed artifacts indicate regular trade with distant lands, including Mesopotamia. Trade gradually dwindled around 2000 B.C. to 1500 B.C., and after this period contact with the outside world ceased. Other Harappan sites have been located over what is today a desert and where irrigation would be difficult, if not impossible.

Several theories have been proposed to explain the decline of the Harappan civilization, including invasions by Aryan barbarians. Wheeler,[48] however, found evidence to indicate that economic decline was brought about by the misuse of the land. The land was denuded by deforestation ultimately breaking the hydrologic cycle and turning the valley into a desert. Bryson and Baerreis[7] also support the misuse theory of deforestation and intensive farming. They proposed that dust was formed by repeated use of land, and by 1500 B.C. the Harappans disappeared from the dust covered desert. In a 1974 issue of the Bulletin of the American Meteorological Society, Ernst[14] shows a satellite picture of the dust layer from drought-stricken Africa that moved over the southwest North Atlantic area. A conspicuous absence of rain producing clouds over Cuba, the Bahamas, and parts of Florida's Atlantic coast was associated with the area in which the dust layer could be observed. He suggests that the dust layer acted as a shield from the energy of the sun and ultimately affected the energy balance that would maintain the differential temperature between land and sea. Dales,[10] a hydrologist, also suggests that climatic changes resulting from deforestation contributed to the migration of the Harappans to other parts of India.

China

From various sources of information, including archeological expeditions, local gazetteers, and diaries, Chu[9] has provided convincing evidence to show that the Yellow River Basin, during the period 3000 B.C. to 1000 B.C. was much wetter and warmer than today. The Yellow River Basin provided a resource area for food-producing civilizations. The fertile alluvial soils washed in from higher elevations by rains, together with abundant precipitation, were ideal for agriculture. Climatic changes, however, have altered historical agricultural patterns and have shifted crop boundaries.

Chinese civilization made its appearance in the Hwang Ho (Yellow River) Valley between 3000 and 2500 B.C.[9] Paintings and archeological discoveries of bamboo rat bones have been related to the climatic optimum period in China corresponding to the time prior to 1100 B.C. These paintings and bones indicate that bamboo groves were quite extensive in the Yellow River Basin. Other evidence includes archeological excavation of animal bones of subtropical origins near Sian.[9] Today the northern limit of bamboo is found 3 degrees of latitude further south and coincides with the 0°C January isotherm which passes from Sian to Suchow (Figure 3). Chu estimates that this 3 degree movement southward corresponds to about a 2°C mean annual temperature decrease when compared with temperatures of today. Drought also seems to have affected the change of ruling dynasties in China. Bishop[3] in discussing the development of Chinese civilization noted that ruling dynasties in 1766 B.C., in 1122 B.C., and again in 842 to 771 B.C. followed periods of droughts. This suggests that conflicts among dynasties were, in part, a consequence of decreasing food supply effected by climatic change.

Wheat, millet, and rice were the staple crops for the early Chinese, as they are today. Hermann[18] noted that the northern limit of cultivated rice has moved since the warm moist period, 722 to 481 B.C. In Figure 3, three boundaries of rice are shown. The first millennium after Christianity was a cold period in China (Figure 1), and the northern limit of rice, which is a warm climate crop, moved 6 degrees of latitude southward during that period. These climatic changes forced Chinese agriculturists to alter management practices of available land as rice could not be grown further north. Sakamoto[37] in analyzing the southern boundary limit for growing wheat noted that since 722 to 481 B.C. the

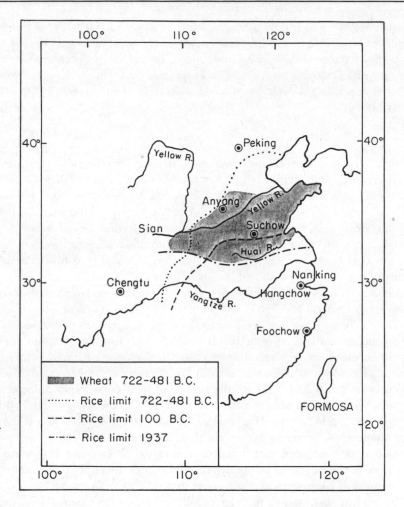

Figure 3. Northern limit of rice culture in China since the first millennium B.C. Darkened area shows the wheat producing area during the period 722 to 481 B.C. Data from Hermann.[18]

boundary had moved southward from the Huai River to the Yangtze River Valley, about 2 degrees of latitude (Figure 3). Also, Chu[9] noted that during the cold 12th and 13th centuries increasing storm tracks further south made the Yangtze Valley cool. It is well known that wheat is a cool climate crop.

Old Silk Road. A controversial site with regard to climatic change is the old silk road of the Tarim Basin in Chinese Turkestan. The drainless basin is bordered by the Tien-Shan mountains to the north, the Kunlun ranges to the south, to the west by the Pamirs, and to the east by the Nan-Shan. The 1500 by 600 mi basin is described as an area void of moisture, human existence, and practically all animals and plant life.[43] Limited life is sustained by irrigation that derives its water from the glacier melt of the Kunlun mountains. The old silk road served as a direct line of communication and trade route between China and the west 2000 years ago. Ideas of domestication of animals and plants passed through this route to the east and eventually to the coastal plain of the Yellow River Basin. Chinese documents and archeological remains indicate that the area was abandoned about the beginning of the third and fourth centuries.[21] From Chinese annals and other historical records Hoyanagi[21] published maps showing the distribution of the remains of the Han dynasty (206 B.C.). He showed that during this period the rivers received more water than today. Indication of later increased aridity is apparent, and during the Chin Dynasty (420 to 265 B.C.) great difficulty was encountered by travelers in passing through the Tarim Basin route. From three expeditions during the period 1900 to 1915 Sir Aurel Stein[43] found artifacts in the Tarim Basin that had withstood 2000 years. This indicates that the local climate was extremely arid then, as it is now, and that it has not substantially changed over the years. However, the flow from the Kulun has diminished. This suggests climatic change in the water bearing mountains.

The Saharas

The Sahara is known for its desert climate. It is difficult to conceive that this desert once was filled with plant, animal, and human life. Yet, the Sahara is one area where archeologists and scholars of other disciplines believe positive evidence exists to show that a fauna flourished about 7000 years ago similar to that found in wetter regions. These evidences are supported with botanical pollen strata analyses, geological and faunal evidence, and also rock drawings.[8] Pollen strata showed signs of lotus, wild olive, cypress, and evergreen oak, all indications of a wetter climate.

In 1933 Archeologist Lhote[29] learned of an area of rich rock paintings in the Sahara and began to search the desert for the now famous rock paintings of Tassili located in a dry plateau in the southeastern corner of Algeria. In 1956 he found hundreds of painted walls of prehistoric art, the oldest one dated 5000 B.C. or earlier. These depict the Sahara full of animal life including giraffe, rhinoceros, hippopotamus, antelope, and others that are found in moister climates. The rock paintings of Tassili, Lhote wrote, "afford us a mass of additional information regarding the types of cultures which have succeeded each other in the desert; culture of hunters armed with clubs and boomerangs; cultures of herdsmen and archers; cultures of warriors whose arms were javelins and who introduced the domestic horse into Africa."

The Tassili frescoes suggest that around 3500 B.C. pastoral invaders entered the Sahara with domesticated animals, including sheep, goats, cattle, and oxen. By about 2000 B.C. the climatic optimum was drawing to a close, and desiccation set in. During that period following the climatic optimum, the atmospheric circulation patterns brought cooler and drier air into Europe and also affected the climate of the Sahara. Vegetative zones in the Sahara shifted as rainfall decreased. Butzer[8] reconstructed a precipitation map comparing the modern amounts with the period 5000 to 3000 B.C. (Figure 4). In the Tassili area precipitation has since decreased by about 60 to 75% and more where once about 100 to 150 mm of rain fell. Butzer also estimates that marginal semiarid vegetation shifted about 100 to 250 km toward the core of the desert.

The frescoes show that near 1200 B.C. invaders with horse driven chariots and javelins and shields moved into the area. Later paintings suggest increasing aridity as hippopotamus and rhinoceros are missing from them. The Sahara, already a desert, was invaded by the Romans about the time of Christianity. The Roman's battle with the environment was a triumph for them as they made the Saharas habitable. They built water storage systems and irrigation channels that carried water over great distances. Farming communities

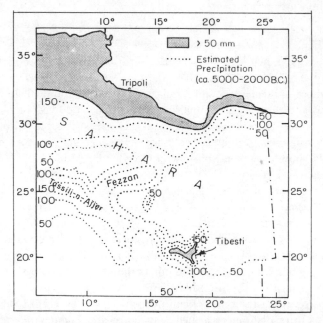

Figure 4. Estimated annual precipitation about 5000 to 2000 B.C. in Central Saharas (dotted line). Darkened area shows modern amount greater than 50 mm. Adapted from Butzer.[8]

flourished, but by 400 A.D. the Roman communities collapsed as bandits con-
quered the settlers. In spite of these apparent achievements of environmental
modification, Lhotc[29] believes that man himself has played a part in causing the
unfavorable climate. Bovill,[5] writing on the encroachment of the Saharas on
Sudan in 1921, cites the case of the desiccation of the Sokoto province in north
Africa. He writes, "first Habe farmers are driven out by drought and diminish-
ing yield, the Fulani or Adarawa move in with their livestock. . . . When graz-
ing becomes too poor for the cattle, the Buzai and Turek with their sheep and
camel are pleased to move in."

Americas

Introduction of agriculture into the southwest developed the village form of
life.[49] These communities flourished, but later declined as they were affected by
environmental factors, notably drought.[25] Like wheat in the old world, maize in
the new world was the sustenance of food growing cultures in the Americas. A
study of the possible origin of this vital crop and its introduction and movement
in the southwest United States has provided an insight to the changing climate
with time. The cultivation of corn probably began in Mexico and Peru. In
Mexico, MacNeish[30] found corncobs that dated about 7000 years. It is believed
that agriculture spread northward and reached western New Mexico about
3500 B.C. Carbon-14 dating of primitive corn found in the Bat Cave placed evidence of farming during this period in western New Mexico. Mangelsdorf[31]
discussed the possible evolution of corn and suggested that environmentally
induced changes have been part of the genetic changes occurring in corn during
its domestication. The area near the Bat Cave, on the shores of a desert lake
bed, was apparently abandoned possibly because of cooling and desiccation.
Pollen strata analyses from the lake bed show large climatic variations[40] and
are consistent with the climatic trend of modern times. Cooler and moister conditions occurred after 3500 B.C. Woodbury[49] notes that following the warm Alithermal, a cool period ensued for 2000 years. A return of warmer temperatures
near the Christian era may have led to the reintroduction of maize from
Mexico into the southwest.[49]

Maize assumed a dominant role in the economy of the southwest around
3000 B.C. and spread rapidly northward to Colorado and Utah and eastward to
the Rio Grande and eventually to Canada.[49] Woodbury[49] also noted that agriculture peaked around 1100 A.D., when it was warmer and drier.
Technological application of irrigation and catchment systems must have been
developed to make more efficient use of the limited water supply. Later in the
12th century, the climate was marked by a rapid change to cooler temperatures. This was followed by drought in the 13th century. Because of the

drought agricultural communities sought more favorable areas at a higher elevation where rainfall supplied the need for profitable farming. At higher elevation, however, the length of the growing season shortened and forced selected management practice, including different planting exposure. Sears,[40] in following maize culture in the Upper Mississippi drainage basin, suggests that civilizations have been affected by climatic changes by shifting the area of optimum food production.

Post Optimum Influences

Influences of Climatic Decline

During this period temperatures gradually decreased. In some regions the period was also drier than the prior epoch. In China documented phenological calendars indicate that during the beginning of the Chou dynasty (1066 to 256 B.C.), the climate was warm enough to grow bamboo extensively in the Yellow River Valley.[9] Toward the end of the 10th century B.C. the climate deteriorated, with some parts of China cooling by as much as 5°C for the month of January and 2° to 3°C annually when compared to the prior warm optimum period. The northern limit of bamboo distribution has since moved southward about 3 degrees of latitude. Compared to the climate today, it was about 1°C cooler and drier during the decline. During this period cool climate affected the staple diet of the people. Glutinous millet, which requires a cool climate for production was the staple crop during this period. During the warmer period from about 900 to about 200 B.C. spiked millet was considered the staple crop in the Yellow River Valley.[9] This indicated that a major shift of crops occurred during this cool period.

The climatic decline was a dry period in Egypt. From Egyptian records, barrages and canals were needed during the rule of Amenemhat III because of the drop of the Nile flood levels. About 1250 B.C. Ramses II constructed the first Suez Canal to link the Mediterranean to the Red Sea. Apparently the drier climate facilitated construction. Further south and west, in the Sahara, the decline from the climatic optimum started about 2000 B.C. With reduced rainfall vegetation zones shifted. Reduced precipitation affected the life of herdsmen and forced them to become nomadic in order to seek water and pasture for their animals.[8] Lhote's[29] rock painting also shows that about 1200 B.C. invaders with horse drawn chariots moved through the area, driving the pastoralists out. Also, the paintings showed absence of the hippopotamus or rhinoceros but still contained animals that can better survive drier climates such as giraffe and antelope.

Influences of "Little Climatic Optimum"

One of the best illustrations of the effect of climatic change on man during this relatively warm period was associated with the colonization of Greenland by the Norsemen. Erik the Red is credited with discovering Greenland. Banished from Norway and Iceland for committing murder, Erik set sail southward from Iceland around 980 A.D. in the sea that was then free of ice. He spent three years exploring the new land and returned to Iceland as a hero to organize another expedition. Two colonies were established in Greenland, and farming sustained the people. Food production was supplemented with hunting, and trade with its mother country, Norway, provided additional supplies. The anonymous *Kings Mirror,* a translation of documents from the early Norse colony, cited successful farming with cattle, sheep, and attempts of growing grain in Greenland.

It is estimated that the average annual temperature during the colonization of Greenland was 2 to 4°C warmer than it is today.[28] Dansgaard and his colleagues[11], analyzed the ice core at Camp Century, Greenland with the use of oxygen-18 isotope concentration as an indicator of past climate. Figure 5 shows that the period about 550 to 1150 A.D. was warmer than the present. If a rough relationship of 1°C increase in temperature corresponding to approximately 0.69% increase in oxygen-18 isotope as reported by Chu[9] is used, the change corresponds to about 1°C to 1.5°C increase. This is only an approximation, as other factors need to be considered before these two variables can be properly related. This warmer period corresponds to the time when the Vikings settled in Greenland and Iceland.[11] Around 1200 A.D. the climate cooled, and drift ice made navigation hazardous in the straits between Greenland and Iceland. By 1300 A.D. commercial trading was severely hampered, and the last recorded voyage between Iceland and Greenland was approximated at 1410 by Mowat.[35]

In 1921 excavation by Hovgaard[20] shed additional light on the Norsemen in Greenland. At Herjolfsfjord (presently Narssarmit), near Cape Farewell in southern Greenland, old records and artifacts including clothing and wooden objects preserved for some 500 years were found in cemeteries. These objects were well preserved in spite of the years. They were found in the soil layers that are presently frozen throughout the year and, therefore, must have been buried when the midsummer heat thawed the soils. Furthermore, plant roots were found engulfing the artifacts in an area where no roots now grow in the permafrost. This finding by Hovgaard provides good evidence of the considerable cooling of that area since the time of the settlement. When the Norsemen settled Greenland, only a trace of Eskimos were found. The migration of Eskimos follows that of the seal. When the seals moved north, so did the

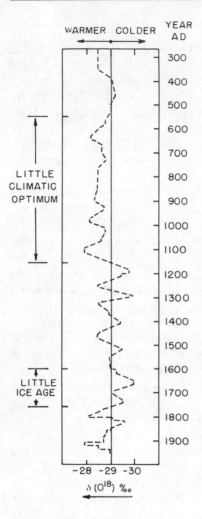

Figure 5. Oxygen-18 isotope concentration on the ice core at Camp Century, Greenland together with approximate climatic stages. Adapted from Dansgaard.[11]

Eskimos. In the cooler 13th century, the Eskimos reappeared in the Norse settlement and advanced further southward until the 14th century. They apparently occupied the settlement and inhabited the entire west coast of Cape Farewell.

Also, Mowatt[35] has observed that in Europe during the little climatic optimum grapes were grown at least 270 miles further north than where they are now grown. During this warm period, 1100 to 1300 A.D., viticulture reached its peak in southern England, and wheat was known to have grown in Kelso[28] (Figure 6). Lamb[28] estimates that in western and middle Europe, during the little optimum, vineyards were grown 4 to 5 degrees of latitude further north and 100 to 200 m higher on the hills. He further estimates that summer temperatures were 1 to 1.5°C warmer prior to 1300 A.D. and concludes that the cool spring and May frost after this period brought an end to most vineyards in England.

Wheat is grown in eastern England today. Historical evidence indicates that problems related to wheat production in this area are associated with excessive rainfall and that a reduction in precipitation in England is not as serious as a reduction in southern Europe.[47]

The little optimum period in China occurred approximately during the

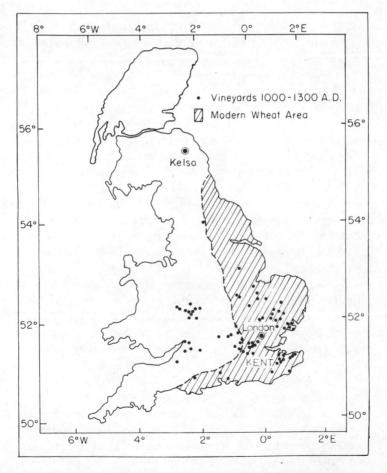

Figure 6. Former vineyards (after Lamb[28]) and modern wheat growing areas of England.

period 500 to 1000 A.D. (Figure 1). During the Tang Dynasty (618 to 907 A.D.) the growing season was apparently longer than today. Chu[9] reports the entries of a book written in 862 A.D. which documented the raising of two crops a year in the region west of Erhai Lake; rice was grown in September and wheat or barley in April. Today, the growing season is too short for the reliability of double cropping.

The warmth and, at times, unusual heat in Europe was responsible for the infestation of locusts in the 9th to 12th centuries.[26] Grasshoppers thrive only when it is warm and dry. These insects spread over vast farming areas and quickly devastated crops. In 873 A.D. the locusts made their presence felt from Germany to Spain, and in the autumn of 1195 A.D. they reached Hungary and Austria.

A period of cooling lasting from the 13th through the 15th century followed the little optimum. The literature is filled with accounts of famine during the cool 14th century.[26, 33, 47] Wheat growing in east and south Scotland was abandoned.[28] Vineyards dwindled after the 14th century. However, successful winemaking has been reported in Kent (Figure 6) in the mid-20th century,[24] when during that period temperatures in England rose 0.5 to 1°C from the "little ice age."

In Iceland economic life was drastically affected during the first half of the 14th century. The economic centers moved from the interior where agriculture played a dominate role to the coastline where fishing became the main occupation.[47]

The little optimum also left its mark in the southwestern United States. Tree-ring analyses, archeological discoveries, and pollen analyses, all seem to indicate the encroaching dryness during the period. The "great drought of the southwest" lasted from about 1276 to 1299 A.D. Indian agriculture in New Mexico and Arizona[49] also flourished in the little optimum between 700 and 1200 A.D. Maize, gourds, and beans were grown, but just before the 13th century the Indian villages began to decline. From the 13th to the 16th century, approximately 70% of the cultivated area was abandoned. Even with the application of irrigation and terracing, the parched land was too great a handicap to overcome. Ladurie[26] cites that, "climate contributed to the launching of a tendency which perpetuated itself through purely human causes and effects." Archeologist Dean[12] also contends that deforestation for agricultural land led to uncontrolled erosion of irrigation channels.

The initial occupation of the Rio Grande Valley by agriculturists is estimated to have started in the vicinity of Las Cruces, New Mexico about 1000 A.D.[25] The Pueblans expanded southward during 1100 to 1200 A.D., which placed them at El Cajon, about 145 miles south of El Paso. Unfavorable fluctuations in climate, possibly increasing aridity, are given as a major causative factor for this move. Geological evidences of the alluvial soils have indicated three distinct soil layers, thought to represent periods of increasing moisture and aridity.[25]

Influences of Little Ice Age

The *little ice age* represents an historical period of pronounced cooling around the middle of the 16th century, which continued to about the middle of the 19th century.[6, 28, 32] The entire northern hemisphere was affected by this period of deteriorating weather.

In Iceland written documents during the 18th century describe the despair of villagers as they noted the movement of the glaciers over places where farmhouses and buildings once stood. From 1753 to 1759 hard freezes, heavy snows, and a lengthy winter left practically no fodder for the cattle. Farmland had to be abandoned.[28] Famine followed, and an estimated 10,000 Icelanders either perished or migrated.[42]

In China winter wheat was planted in the latter part of November during the 16th century. During the cooler 17th century, especially during the epoch 1650 to 1700, winter wheat was sown two weeks earlier to escape the cold damage from winter. Also, in China records were kept at Peking from 1724 to 1903. From these records, Chu[9] concludes that the period 1801 to 1850 was warmer than the period 1750 to 1800 and the period 1851 to 1900.

The harvest failures in England during the 1550s and 1590s were disastrous. Famine occurred four consecutive years, 1594 through 1597. Cool springs and summers and excessive rainfall forced farmers to leave their wheat crops in the fields to rot. The cool weather seemed to persist through the 17th century. Feudal wars broke out, turmoil and national unrest followed.

Utterstrom[47] cites several examples to show that fluctuations of climate were important to the economic and social history of Sweden. He provides examples from the 16th and 17th centuries, which markedly affected most of Europe and the Scandanavian countries as well. In western Sweden the parish register contained a document describing in detail crop failures and desperation of people from famine. The parishioner writes, "People ground and chopped many unsuitable things into bread; such as mash, chaff, bark, buds, nettles, leaves, hay, straw, peat moss, nutshells, etc. . . . people, men and women were compelled in their hunger to stealing. . . ."[47]

The bubonic plague epidemic of the 17th century is considered in light of crop failures. Dr. F. Hirst,[19] writing on the conquest of the plague, believed that the main reason rats migrated was because of hunger. The rat fleas, carriers of the plague, sought other hosts—people—because the rats died or fled elsewhere. When the people noticed that rats were dying, they left their homes where this relationship was known. The bubonic plague is not considered the only factor in the decline of the population during that time. Among other factors was the climate, which affected not only the people but also the rats and the fleas as well. Utterstrom[47] believes that the years of famine and plague during the 16th and 17th centuries were important turning points in the development of Swedish population.

Summary

There can be little doubt that past climatic changes have altered human lives throughout the world. Evidence by researchers in varied environmentally related disciplines have been presented in this review. Only a small portion of these, however, are discussed. Nevertheless, they point strongly to climate as a contributory factor in shaping the economic, cultural, and social history of many civilizations. Hustich[23] has aptly discussed the mechanism of response to historical climatic changes. He writes, "climatic changes influence the balance within a community and different members react differently to it. This, in turn, leads to a displacement of population boundaries."

It is noteworthy that a gradual decline of many past civilizations was related to the availability of water for food production. In an attempt to produce a greater quantity of food more effectively for the rapidly expanding population forests were cleared, and when water became scarce, irrigation systems were developed. In many areas the abuse of irrigation led to a gradual deterioration of the once fertile lands. Therefore, man himself contributed to the alteration of the energy balance and short circuited the hydrologic cycle. On the other hand, the reasons for large scale global atmospheric changes over a period of time remain obscure, although many hypotheses have been proposed.

Analyses of the response of man to past climatic changes have revealed exploration of new land, cultural changes, forced migration to more favorable areas, and economic chaos followed by conflicts, famine, and death by the thousands. Even in the 20th century these responses are familiar. History has also revealed that man has adapted himself to these changes with new technological inventions and knowledge. How far he can go with this adaptation remains to be seen.

References

1. Adams, R. M., "Agriculture and Urban Life in Early Southwestern Iran," *Science*, Vol. 136, No. 3511 (April 13, 1962), pp. 109–122.
2. Arakawa, H., "Climatic Change as Revealed by the Blooming Dates of Cherry Blossoms of Kyoto," *Journal of Meteorology*, Vol. 13 (1956), pp. 599–602.

3. Bishop, C. W., "The Geographical Factor in the Development of the Chinese Civilization," *The Geographical Review,* Vol. 12 (1922), pp. 19–21.

4. Bonatti, E., "North Mediterranean Climate During the Last Wurm Glaciation," *Nature,* Vol. 209, No. 5027 (March 5, 1966), pp. 984–985.

5. Bovill, W. E., "The Encroachment of the Sahara on the Sudan," *Geographical Review,* Vol. 11 (1922), pp. 622–623.

6. Brooks, C. E. P., *Climate Through the Ages,* New York: McGraw-Hill, 1949.

7. Bryson, R. A. and Baerreis, D. A., "Possibilities of Major Climatic Modification and Their Implications: Northwest India, A Case for Study," *Bulletin American Meteorological Society,* Vol. 48, No. 3 (1967), pp. 136–142.

8. Butzer, K., *Environment and Archeology,* Chicago: Aldine, 1964.

9. Chu, C., "A Preliminary Study of the Climatic Fluctuations During the Last 5,000 Years in China," *Scientia Sinica,* Vol. 16, No. 2 (May 1973), pp. 226–256.

10. Dales, G. F., "The Decline of the Harappans," *Scientific American,* Vol. 214, No. 5 (May 1966).

11. Dansgaard, W., Johnsen, S. J., Miller, J., and Langway, C., "One Thousand Centuries of Climatic Record from Camp Century on the Greenland Ice Sheet," *Science,* Vol. 166 (Oct. 1969), pp. 377–381.

12. Dean, J., *Chronological Analysis of the Tsegi Phase Site in North-East Arizona,* Unpublished Thesis, Tucson: University of Arizona, 1967.

13. Douglass, A. E., Carnegie Institute of Washington Publication No. 289, I (1919), II (1928), III (1936).

14. Ernst, J. A., "African Dust Layer Sweeps into the Southwest North Atlantic Area," *Bulletin American Meteorological Society,* Vol. 55, No. 11 (November 1974), p. 1352.

15. Fritts, II. C., "Growth Rings of Trees: A Physiological Basis for the Correlation with Climate," in R. H. Shaw, Ed., *Ground Level Climatology,* Washington, D.C.: Association for the Advancement of Science, 1967, pp. 45–65.

16. Gentilli, J., "Quaternary Climates of the Australian Region," *New York Academy of Sciences Annals,* Vol. 95 (1966), pp. 465–501.

17. Helbaek, H., "Domestication of Food Plants in the Old World," *Science,* Vol. 130 (1959), pp. 365–373.

18. Hermann, H., *An Historical Atlas of China,* New ed., Chicago: Aldine, 1966.

19. Hirst, L. F., *The Conquest of Plague,* London: Oxford University Press, 1953.

20. Hovgaard, H., "The Norsemen in Greenland: Recent Discoveries at Herjolfsnes," *The Geographical Review,* Vol. 15, (1925), pp. 605–606.

21. Hoyanagi, M., "Sand-Buried Ruins and Shrinkage of Rivers Along the Old Silk Road in the Tarim Basin," *Japanese Progress in Climatology,* Tokyo 2, 76–86, 1965.

22. Huntington, E., *Civilization and Climate,* New Haven: Yale University Press, 1924.

23. Hustich, I., "On Variations in Climate in Crop of Cereals and in Growth of Pine in Northern Finland 1890–1939," *Fennia,* Vol. 70 (1947), p. 2.

24. Hyams, E., *An Englishman's Garden,* London-Southhampton: Camelot, 1971.

25. Kelley, J. C., "Factors Involved in the Abandonement of Certain Peripheral Southwestern Settlements," *American Anthropology,* Vol. 54 (1952), pp. 356–387.

26. Ladurie, E. L. R., *Times of Feast, Times of Famine,* Garden City, New York: Doubleday, 1971.

27. Lamb, H. H., "Britain's Changing Climate," in C. P. Johnson and L. P. Smith, Ed., *Biological Significance of Climatic Changes in Britain,* London-New York: Academic Press, 1965.

28. Lamb, H. H., *The Changing Climate,* London: Methuen, 1966.

29. Lhote, H., *The Search for the Tassili Frescoes,* New York: E. P. Dutton, 1959.

30. MacNeish, R. S., "Preliminary Archeological Investigations in the Sierra de Tamaulipas, Mexico" *Transactions of the American Philosophical Society (n.s.),* Vol. 48, No. 6 (1958), pp. 1–210.

31. Mangelsdorf, P. C., "Ancestor of Corn," *Science,* Vol. 128 (1958), pp. 1313–1320.

32. Manley, G., "The Range of Variation of the British Climate," *Geographical Journal* Vol. 117 (1951), pp. 43–68.

33. Manley, G., "The Revival of Climatic Determinism," *Geographical Review,* Vol. 48, No. 1 (1958), pp. 48–105.

34. Manley, G., "The Problem of the Climatic Optimum: the Contribution of Glaciology," in *Proceedings of the International Symposium on World Climate 8000 to 0 B.C.,* Royal Meteorological Society, London, 1966, pp. 34–39.

35. Mowat, F., *Westviking,* New York: Minerva, 1968.

36. Oliver, J. E., *Climate and Man's Environment,* New York: John Wiley, 1973.

37. Sakamoto, C., "Effect of Climate on Non-Domestic Wheat Production," in A. J. Broderick and T. M. Hard, Ed., *Proceedings of the Third Conference on the Climatic Impact Assessment Program,* DOT Transportation System Center, Cambridge, Massachusetts, 1974, pp. 539–549.

38. Schulman, E., "Tree-Ring Indices of Rainfall, Temperature and River Flow," in T. Malone, Ed., *Compendium of Meteorology,* Boston: The American Meteorological Society, 1951, p. 1024.

39. Schwarzbach, M., *Climates of the Past,* Princeton: Van Nostrand, 1963.

40. Sears, P., "Climate and Civilization," in H. Shapley, Ed., *Climatic Change,* Cambridge: Harvard University Press, 1965.

41. Shen, W., "Changes in China's Climate," *Bulletin American Meteorological Society,* Vol. 55, No. 11 (1974), pp. 1347–1350.

42. Silverberg, R., *The Challenge of Climate,* New Haven: Yale University Press, 1969.

43. Stein, A., "Innermost Asia: Its Geography as a Factor in History," *Nature,* Vol. 114 (1924), pp. 805–806.

44. Taylor, G. C., "Water, History and the Indus Plain," *Natural History,* Vol. 74, No. 5 (May 1965), pp. 40–49.

45. Taylor, S. A. and Ashcroft, G. L., *Physical Edaphology,* San Francisco: W. H. Freeman, 1971.

46. Titow, J., "Evidences of Weather in the Account Rolls of the Bishopric of Winchester 1209–1350," *Economic History Review,* 2nd series, Vol. 12 (1960), pp. 360–407.

47. Utterstrom, G., "Climatic Fluctuations and Population Problems in Early Modern History," *Scandinavian Economic History,* Vol. 3, No. 1 (1955), pp. 3–47.

48. Wheeler, M., *The Indus Civilization,* 3rd ed., Massachusetts: Cambridge University Press, 1968.

49. Woodbury, R. B., "Climatic Changes and Prehistoric Agriculture in the Southwestern United States," *New York Academy of Science Annals,* Vol. 95 (1961), pp. 705–709.

2

Man's Impact on Climate

HANS PANOFSKY
Evan Pugh Research Professor of Atmospheric Sciences
The Pennsylvania State University
University Park, Pennsylvania

Climate is defined as the statistics of meteorological variables. The most important of these are temperature and precipitation, but a complete discussion of climate would include such factors as wind, humidity, cloudiness, and other weather indicators.

Climate involves average values, usually computed over 30 years. The choice of this period is arbitrary, and is largely a result of availability of weather data. In addition, climatology involves statistics of weather variability, for example, the variability from day to day or throughout the year; and it involves the statistics of extremes such as the probability of extreme temperatures or winds at a given site.

For the purpose of this chapter, we consider three categories of climate, classified according to the area affected: *local climate, regional climate,* and *global climate.* Local climate represents the special statistics of weather elements in areas up to a few miles square. In such areas, man-made influences have been recognized for a long time, and many can be estimated quantitatively. For example, urbanization produces "heat islands," slow winds, relatively strong air turbulence, and increased precipitation in the immediate area and environment of the city. Irrigation and agricultural activity clearly influence temperature and humidity in the immediate neighborhood.

Regional climate includes areas of the size of individual states in the United States, or of small countries. Here, man's influences are understood less quantitatively, but certainly occur and will increase as many of man's activities responsible for such changes increase. To illustrate, local deserts are apparently formed as a consequence of overgrazing in western India; and acid rain is falling over many parts of the world as a result of SO_2 being released into the atmosphere by industry.

Changes in global climate have occurred over the whole period of the earth's history. Major variations in temperature and ice cover have taken place on scales of 100 million, and hundreds of thousands of years; somewhat smaller changes have occurred on scales of centuries and thousands of years since the last ice age. These recent changes have not been explained. But it is quite clear that global climatic changes in the last centuries of increasing human activity have not been different in character from the changes in earlier periods. There is no evidence that the global climate so far has been affected by human activity.

At the same time, there is no agreement on what caused the climatic changes on the various time scales. On the very longest time scale, continental drift and mountain building must have certainly had effects. The advance and recession of the major ice sheets in the last million years has been ascribed to many possible causes: changes in the sun's heat, changes in the earth's orbit, changes in suspended particles due to varying volcanism, and alternate increases and decreases of CO_2 in the atmosphere—to mention only the most prominent theories. Most of these mechanisms span too long a time scale to explain the most recent climatic changes. No generally accepted theory for these changes exists; hence, no predictions can be made as to how climate would behave in the next century or so in the absence of human interference.

For example, between 1880 and 1940 the global mean temperature increased at the rate of about 1°C per century; there is some evidence that it has since decreased at nearly the same rate. In polar regions, natural climatic changes are always larger than in the tropics; and evidence of recent cooling is especially strong in the arctic. Some scientists have ascribed this recent cooling to increased loading of industrial particles in the atmosphere; but it could have been due to unknown "natural" causes just as well.

Thus, the observed record is not much help in detecting possibly disastrous climatic changes; by the time they exceed the threshold of "natural" variability, the effect may be out of control. Instead, it is necessary to review all facets of human activity and try to predict the kind of global changes likely to ensue.

The potential seriousness of the problem of anthropogenic climatic changes led the Massachusetts Institute of Technology to organize a national workshop on this subject in 1970 and an international workshop in Stockholm in 1971, both under the leadership of Prof. Carroll Wilson. The results of both workshops have been published by MIT Press. Much of the material in this chapter is based upon the second volume, "Report of the Study of Man's Impact on Climate (SMIC)"[7]. An even more detailed discussion of current ideas on anthropogenic climatic change is contained in Monograph 4 of Project CIAP, U.S. Department of Commerce.

This chapter concentrates on possible man-made climatic changes on the global scale, mentioning regional changes occasionally; it ignores local changes.

Climatic Models

The natural climate is controlled by many factors: the distribution of incoming solar radiation; certain properties of the atmosphere; and some ground characteristics. Of course, only the distribution of sunshine at the top of the atmosphere cannot be influenced by man. But both the atmospheric and ground conditions can be and have been changed by man in many ways. Thus, particles and gases are being discharged into the atmosphere at an increasing rate (sometimes with gases changing into particles). These substances affect the transmission of sunlight through the atmosphere, either directly, or by causing alteration of cloud patterns; and the absorption and emission of infrared radiation also depend on the distribution of gases, particles, and clouds.

Changes in human activity alter the reflectivity of the ground, and can affect the water supply, which in turn modifies atmospheric conditions, particularly cloudiness and precipitation.

All these processes affect the distribution of temperature, but in a most complex manner. For example, when the temperature is changed, infrared radiation is affected, which in turn, affects the temperature distribution. Also, as the temperature varies, winds are changed, which redistributes the pollutants, giving rise to altered patterns of temperature and precipitation. Or, most important, a cooling trend causes the polar ice sheets to increase. Ice is an excellent reflector of sunlight, so that less sunlight remains in the earth-atmosphere system—leading to additional cooling, and additional spreading of the ice sheet. This process is called *albedo instability* (albedo stands for reflecting power) and has led some investigators to suggest that a decrease in solar radiation by only a few percent will rapidly produce an ice sheet covering the whole earth. The fact that this has never happened suggests that this conclusion must have been based on too simple a model, and that many other processes must have been triggered by the expanding ice sheet which were not taken into account.

It is clear from the foregoing that the interdependence of meteorological and terrain variables makes a prediction of the impact of human activity on climate exceedingly complex. Such predictions are carried out by means of climate models. These range in complexity from simple global-average models to fully three-dimensional hydrodynamic models.

Global-average models proceed from statistical relationships between radiation and temperature, and infer from these how a change in the visible or infrared radiation will affect the temperature. Since only global averages are considered, winds are ignored. The only physical restraint on such a model is that energy must balance. Models of this type have been constructed for various ground characteristics, and various contaminants. One important result, which still is quite uncertain, is that the reduction of sunlight by 1% due to absorption by a contaminant, leads to a temperature decrease of about 2°C. But these models make important simplifications. For example, neither clouds nor the ocean are handled realistically.

Changes in cloudiness and changes in the ocean temperatures, produce special problems. Clouds have a rather complex effect on climate. On the one hand, they reflect sunlight, causing atmospheric cooling; on the other, they generally emit less infrared radiation into space than would have been emitted without them, so that increased cloudiness would lead to warming. This effect would dominate at night (no sunlight) or at high latitudes. But clouds appear in all kinds of shapes at various elevations; so far neither the simple global-average nor the complex hydrodynamic models contain satisfactory estimates of the role of changing cloudiness in modifying the climate.

Although the exact role of the oceans in the processes of climatic changes is not known, it seems reasonable to suppose that, with their large heat capacity, they will decrease temperature changes that would occur in the atmosphere without it, or, at least, slow the changes down for centuries.

Complete hydrodynamic models integrate simultaneously all the basic equations governing the atmosphere, and sometimes the oceans, as well, by numerical techniques. Thus, in addition to the heat equation and equations of state, the fluids must obey Newton's Laws and conservation of mass for the air in the atmosphere and the water in the oceans. Also, individual substances contained in the fluids must be conserved such as water vapor and ozone* in the air, and salts in the water. In the integration, the properties of the ground appear in the form of internal boundary conditions.

Typically, climate models have been used to explain the existing climate, not to predict climatic changes resulting from changes in the boundary conditions, or of physical properties of the atmosphere. In explaining the observed global temperature distribution and the observed General Circulation of the Atmosphere, many of the hydrodynamic models have been quite successful. Some models also represent the precipitation patterns accurately! But in hydrodynamic models clouds are not predicted properly; and air-sea exchanges are not handled realistically.

Such models generally begin with an observed weather pattern or with a simple atmospheric state as initial condition; numerical integrations are then executed for many model days, sometimes for over a year. Then the statistical properties of the model are compared with the statistical properties of actual weather data. But there are strong variations from day to day, and year to year, both in the model and in the weather. If we now change the initial conditions to reflect the effect of some human activity, for example, add some parti-

* Except for chemical changes.

cles to the model, the detailed fluctuations in the model will change, but the statistics, such as the January average, may not change "significantly."

As a result of the imperfect representation of the physics, and the statistical fluctuations in the "models," then, it has not been possible to make complete quantitative estimates of many anthropogenic climatic change; only orders of magnitude can so far be estimated. Some of those are indicated in the following discussion. Only in the case of CO_2 pollution, have reliable quantitative calculations been made so far.

Byproducts of Human Activity Affecting Climate

Agriculture

When agricultural land is reclaimed from forest on a large scale, the terrain is changed in many ways. First, farm land is not able to hold its water as well as forest. More of the sun's heat is conducted directly into the atmosphere, less heat is used for evaporation. Hence, the atmosphere will be warmer over the land and less cloudy. The surface temperature will be greater. New circulations will be set up between the farming area and the surrounding area. No model of such changes exists; presumably the climate is affected regionally, unless the reclaimed area is huge.

Irrigation of an otherwise dry area produces opposite changes; so do man-made lakes (for which some regional models do exist).

Further, changes of the surface usually produce changes in the albedo (reflecting power) of the surfaces. Particularly in a snow-covered area in winter, the reflecting power of agricultural land is much greater than that of forests, so that much heat is lost to space. Such changes must be considered in modeling any widespread terrain modifications.

Particularly serious consequences in certain regions may have been brought about by overgrazing. Grasses are gradually destroyed and replaced by deserts. With strong winds, sand particles are carried into the air. These radiate in the infrared, cool, and cause the air to sink. Precipitation cannot form in sinking air, and the desert remains. This mechanism (described by Bryson[1]) is quite controversial; but in places such as northwestern India the desert advances, and the climate changes correspondingly.

Agriculture also contributes to the global atmospheric particle load, though probably not a great deal. As we demonstrate, accumulation of particles in the atmosphere is one of the major processes capable of global climatic change.

Urbanization

Cities release more heat to the atmosphere than does the surrounding countryside. As mentioned above, this has important local consequences such as the formation of heat islands. Averaged over the globe, this heat is a very small fraction of the sun's heat. However, as cities grow and complexes such as megalopolis are created (from Boston to Washington), larger regions will be affected by this "heat pollution"; almost certainly, when the general wind patterns are weak, such city groups will set up their own regional circulations and temperature patterns. It is unlikely that such patterns will be of global significance.

Cities are sources of many types of pollution. Many types of particles are formed, especially where much of the refuse is incinerated as in New York. Sulfur dioxide is emitted from heating plants and converted into sulfates. Automobiles (concentrated in cities) are responsible for smog particles. Among the gases affecting climatic change, CO_2 is emitted wherever fossil fuels are burned. Again, such pollution sources are concentrated in cities.

Industrial Activity

It is, of course, somewhat artificial to separate effects of industrial activity from effects of urbanization, because much industry is located in cities. But industry, no matter where it is found, is a major source of SO_2 (and therefore sulfate particles), CO_2, and unburned carbon particles, as well as other gases and particles characteristic of particular industrial processes.

Transportation

Motor vehicles are blamed for over 50% of the pollution in the United States. They are sources of CO, CO_2, SO_2, NO, NO_2 and hydrocarbons.[2] NO_2, under the action of sunlight, releases an oxygen atom that attaches to an oxygen molecule to form ozone, O_3. This combines with the hydrocarbons to form complex molecules known as *smog*. Smog decreases the local visibility, increases the temperature, and adds to the atmospheric burden of particles.

Aircraft emits the same pollutants as automobiles. In addition, we must consider contamination by water vapor, which is unimportant at ground level but could lead to climatic changes when emitted at high levels, as discussed later.

Tropospheric Pollution and Climatic Change

Carbon Dioxide (CO_2)

The troposphere is the lowest layer of the atmosphere, extending from the ground to about 6 mi at the poles and 10 mi at the equator. In the troposphere the temperature generally decreases upward. Precipitation is almost entirely confined to the troposphere. Air in the troposphere is generally well mixed. If contaminants are emitted into the troposphere, they are not likely to remain there much more than a week, because they are removed by precipitation and deposition at the surface. Therefore, relatively large amounts of pollution can enter the troposphere (compared to higher layers) before the surface climate is affected.

Carbon dioxide is one of the most important contaminants deposited in the atmosphere. Burning of any fossil fuel results in the production of this gas. Good estimates exist for the total amount of CO_2 emitted into the atmosphere. Also, the concentration of CO_2 is being measured accurately at several stations. Apparently CO_2 is so stable that its concentration is the same almost everywhere in the troposphere and stratosphere, except very close to the ground.

Figure 1 shows that the fraction of CO_2 in the air (by volume) has increased from about 312 ppm in 1958 to 322 ppm in 1970, according to SMIC.[7]

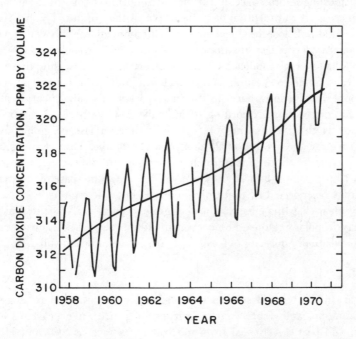

Figure 1. Mean monthly values of CO_2 concentration at Mauna Loa, Hawaii, for the period 1958 to 1971. Sources: For 1958 to 1963, Pales and Keeling, 1965. For 1963 to 1971, Keeling and Bainbridge (unpublished), 1971. The source of the best-fit third-degree polynomial curve for the period 1958 to 1969 is Cotton, National Oceanic and Atmospheric Administration, United States (NOAA). Taken from SMIC.[7]

Recently, Keeling (unpublished) has shown that these values are 4 ppm too low. Until 1970 a little less than half of the CO_2 emitted into the atmosphere remained there. The remainder must either have been dissolved in the oceans, or fixed in the biosphere to grow more vegetation, or both. On the basis of this increased concentration, Machta[4] has constructed a model for the change of CO_2 concentration with time, suggesting a concentration of 375 to 380 ppm by the year 2000.

However, a personal communication from Machta notes that since 1970 the atmosphere has retained 65% of the CO_2 emitted into it instead of the earlier 45%. The reason for this change is unknown. If this large fraction remains unchanged, the concentration in the year 2000 could easily reach 400 ppm; but this prediction is uncertain because the mechanism for removal of CO_2 is so poorly understood.

CO_2 is transparent to sunlight, but has a strong absorption band near 15 μm in the infrared. This means that an increase of CO_2 will increase radiation from the atmosphere into space and from the atmosphere toward the ground. It will also increase the absorption of radiation from the ground. The net result is that the ground will lose less heat, and the air immediately above the ground will be warmed; in contrast, the upper atmosphere will be cooled.

Many global-average models have been constructed for assessing these effects quantitatively. Some of these are quite elaborate, allowing for the presence of water vapor, which also radiates in the infrared with its concentration affected by warming and cooling. According to Manabe and Wetherald,[5] the best estimate so far is that a doubling of CO_2 will cause a rise of global average temperature near the ground of 2°C. But arctic temperature changes can be four times as large. Since it is difficult to see how the CO_2 concentration can increase by much more than 25% by the year 2000, an increase of temperature of 0.5°C may be expected by that time, and 2°C in arctic regions.

It seems clear that this effect is not negligible, even as early as the year 2000, particularly in polar regions. The carbon dioxide concentration must be monitored, and eventually burning of fossil-fuels may have to be curtailed.

Particles

There are many natural sources for particles in the atmosphere: sea spray, blowing sand, volcanic debris, particles from forest fires, hydrocarbons exuded by plants, and particles formed from natural gases such as SO_2 and H_2S. To this natural burden, man's many activities add various types of additional particles.

The fraction of atmospheric particles contributed by man is very uncertain; estimates average about 20%, but the uncertainty of this estimate is almost as large as the estimate itself. By the year 2000, the estimates are even more uncertain; it is quite possible that the anthropogenic particulate loading may approach the natural loading.

Particles in the atmosphere can interfere with normal climatic processes in many ways. They reflect sunlight; if the particles are light, above a dark surface, sunlight will be lost that would be retained in their absence; if the particles are dark above a light surface, more sunlight will be retained. Particles can absorb sunlight, heating the layers in which they are embedded. Particles also emit and absorb infrared radiation, modifying the infrared radiation normally emitted and absorbed by the air.

More indirectly, particles can influence the heat budget by acting as nuclei for clouds that would not have occurred otherwise; or particles can modify the optical properties of existing clouds.

All the effects of particles depend on their color, their size distribution, their shapes, and their numbers. Little is known about any of these characteristics. We cannot be sure whether additional particles would cause heating or cooling of the air near the ground; very uncertain estimates suggest that cooling would dominate.

Thus, we are very far from any quantitative estimate of temperature changes brought about by man-made particles. There is not even general agreement whether atmospheric particle loads have increased far from human activity. Unlike gases like CO_2, particles have their own fall velocities, and are only poorly mixed over the globe. A huge number of stations is required for a detailed particle census. There is no question that the particulate loading has increased and is increasing in and near industrial centers, and that the concentration of smog particles is increasing near many cities; further, these increases certainly have effects on the local climate of these regions; but a global increase of particles has not been well documented; for example, Dyer[3] has produced evidence that in some remote regions, the transparency of the atmosphere has not deteriorated, suggesting that the particle loading has not increased in recent years.

In spite of all this uncertainty and controversy, it is clear particles could have an important impact on climate in the next 50 years, and additional monitoring is needed, coupled with research into the physical properties of particles. Since the most important source of anthropogenic particles is the conversion from SO_2 to sulfates, desulfurization of fossil fuels takes on additional urgency.

Water Vapor

Industrial cooling towers and cooling ponds add to the atmosphere water vapor that would not have been there otherwise. In certain atmospheric conditions, steam plumes from cooling towers have been followed for many miles downwind, sometimes causing fog and ice on highways. However, problems of this type appear to be largely local, or at most, regional.

Conventional aircraft also emit water vapor into the troposphere. When the humidity is high enough, this excess vapor condenses in the form of vapor trails. Sometimes, these vapor trails spread laterally and appear as cirrus clouds.

There is some evidence, particularly at Denver and Salt Lake City, that cirrus clouds have increased significantly in about the same manner as the consumption of jet fuel has increased (see Figure 2, taken from SMIC[7]). The evidence is controversial, however, because middle clouds seemed to decrease at

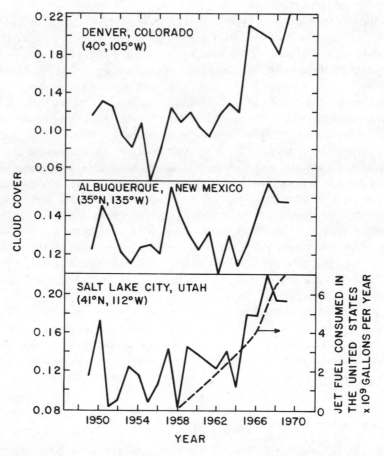

Figure 2. Average annual high cloud cover with zero low or middle clouds for Denver, Colorado, Albuquerque, New Mexico, and Salt Lake City, Utah. Original taken from SMIC.[7]

the same time, perhaps because the observer classified as cirrus clouds he previously would have classified as middle clouds.

Calculations and measurements so far suggest that the effect of jet-produced cirrus clouds on temperature near the ground is negligible; however, locally, strong decreases of sunlight have been measured.

Stratospheric Pollution and Climatic Change

The stratosphere extends from the top of the troposphere up to 30 mi. In it, the temperature generally is nearly constant with height or increases with height. As a result, vertical mixing is slow. Any material deposited in the middle or the top of the stratosphere will remain there for a long time. Also, its density is less than the density of the troposphere; the density in the center of the stratosphere is only about 0.01 the density at the ground. For these two reasons, a given mass of pollutant in the stratosphere can have much more serious consequences than the same mass in the troposphere.

The principal stratospheric polluters considered so far are aircraft, particularly supersonic aircraft (SSTs), should they proliferate. Conventional jets usually fly in the troposphere; and when they penetrate into the stratosphere, they only fly in its lowest levels, from where the pollution rapidly spreads down into the troposphere and is removed. Even here the possibility exists that the next generation of conventional jets will fly up to 10,000 ft higher than present jets; if more than several hundred of such jets are built their effect cannot be neglected, unless the engines are modified to curtail emissions.

Aircraft emit water vapor, carbon dioxide, sulfur dioxide, oxides of nitrogen and small amounts of soot. Of these sulphur oxides, nitrogen oxides, and soot can be controlled; water vapor and carbon dioxide cannot.

Another potential source of stratospheric pollution is the production near the ground of extremely stable gases. Such gases will gradually seep into the stratosphere and eventually be broken up by ultraviolet light. If the resultant species are strongly reactive, serious problems can result. So far, the gases of this type considered are the chlorofluoromethanes (CFMs), also called Freons, although Freon is just a brand name of some of the best-known CFMs. These gases are used in aerosol sprays and as refrigerants. Measurements suggest that all of the CFMs produced so far are still in the atmosphere—there seem to be no important natural loss processes. They have just been detected in the stratosphere.

The main harmful product of dissociated CFMs is chlorine; the actions of this gas in the stratosphere are described later. Other sources of chlorine are chlorinated water bodies and some cleaning fluids; here, too, chlorine may eventually reach the stratosphere.

We shall now consider the action of each pollutant separately, and consider to what extent it can influence climate.

Carbon Dioxide

The action of carbon dioxide has already been discussed in connection with tropospheric pollution—it absorbs and radiates strongly in the infrared and thus alters the atmospheric heat exchange. However, the additional CO_2 possibly added by stratospheric jets is so small compared to the CO_2 already there, or compared to tropospheric CO_2 pollution, that it is completely negligible.

Water Vapor

The stratosphere is quite dry, containing, on the average, only four molecules of H_2O per 10^6 molecules of air. Calculations suggest that a sizeable fleet of SSTs (such as may possibly be constructed by the year 2000), will add about 10% to this concentration. So, unless stratospheric water vapor now influences climate strongly, it is unlikely that this small addition will be important.

Three possible influences of H_2O have been considered: its influence on the heat budget because it, like CO_2, is active in the infrared; formation of clouds from the H_2O; and its effect on the ozone in the stratosphere.

Emission of infrared light by the additional water vapor in the stratosphere will cool the stratosphere itself, and will warm the lower layers. All these effects are estimated to produce temperature changes of the order of 0.1°C, which can be neglected.

Clouds are rare in the stratosphere. The reason is that condensation of the small amount of stratospheric H_2O would require a temperature of about −87°C or less. Such temperatures occur only in polar regions in winter. There we sometimes see mother-of-pearl clouds (also called nacreous clouds). Stanford and Davis[6] have recently collected all reports of such clouds, and found about 150. Of course, the actual cloud amount may be larger, because such clouds probably occur largely in the polar night. Further, they are more common in the antarctic than in the arctic, because antarctic temperatures are lower; and observations in the antarctic have only begun quite recently.

Since such clouds occur only in polar regions in winter, they cannot have any important influence on visible light. But they may affect outgoing infrared radiation if they emit amounts of radiation significantly different from those that would be emitted in their absence. But, in that case, they should be seen from weather satellites with infrared sensors. But nacreous clouds have never been detected from satellites. The reason is probably a combination of factors:

first, the clouds are likely to be thin and therefore not good radiators; and second, particularly in the antarctic, they would probably radiate at about the same temperature as their environment. Thus, stratospheric clouds do not seem to affect climate significantly. Hence, it is extremely unlikely that an additional 10% of water vapor will be of any importance.

Water vapor can attack atmospheric ozone; but oxides of nitrogen affect the ozone even more. Water vapor actually removes some of the oxides of nitrogen by forming nitric acid, so that, in this indirect way, it actually preserves some ozone which without it would have been destroyed.

We can tentatively conclude, therefore, that stratospheric water vapor pollution is not a serious problem.

Sulfur Dioxide

Sulfur dioxide pollution may affect climate because most of it is rapidly transformed into sulfate particles, which interfere with radiation in a number of ways (as already discussed in connection with tropospheric pollution). Of course, the exact effect depends on the size distribution and optical properties of the particles, which are not well known. Further, we have to add these uncertainties to the difficulty of converting changes in radiation into temperature changes. Still, some estimates of global average temperature changes due to possible fleets of aircraft have been made; the reduction of sunlight is expected to be the main effect, and there is expected to be a possible global temperature decrease of order of $0.3°C$, with larger cooling at high latitudes.

Although such estimates are extremely uncertain, it is thought to be prudent to radically decrease the SO_2 problem by reducing the sulphur concentration in the jet fuel for SSTs from the current 500 ppm to about 25 ppm. This is a kind of climate insurance, costing only about one cent per gallon of jet fuel, or about two dollars per passenger ticket.

Oxides of Nitrogen

NO_2 absorbs sunlight quite efficiently so that even a small amount of this gas emitted into the stratosphere can cool the ground temperature significantly. In fact, cooling due to stratospheric SST NO_2 pollution has been estimated to be of comparable importance to the cooling due to sulphate pollution.

In addition to this direct effect on climate, oxides of nitrogen have many indirect effects by removing some of the stratospheric ozone. Quite a small amount of these gases can dissociate a very much larger amount of ozone. This is because NO acts like a catalyst; the formation of two ordinary oxygen molecules O_2 from a combination of an oxygen atom O and an ozone molecule O_3 is much speeded up by the presence of NO; and NO is not destroyed in this chemical process. The complete chemistry of stratospheric minor constituents is very complex, involving not only ozone, oxygen, and NO, but also many other gases and radicals such as NO_2, H_2O, and OH. About 50 chemical equations are estimated to be necessary to solve the problem of how much ozone is destroyed by a given injection of contaminants.

In addition to the chemistry, atmospheric transport has to be taken into account; often changes due to atmospheric motions are about equally fast as chemical changes. No really complete model of all these processes exists. Either, the complete chemistry is combined with greatly simplified atmospheric motion, or a complex three-dimensional atmospheric motion is combined with incomplete chemistry, or some compromise is attempted. Still, estimates of ozone changes due to SST emissions are not extremely controversial; the higher the flight levels, the better the agreement of the different estimates. For example, it is estimated that the maximum likely SST fleet will reduce stratospheric ozone in the Northern Hemisphere by about 10% in 1990 and by more than 30% in 2005. These estimates are computed from a model that yields values on the high side, but effects less than two or three times as small are unlikely.

The best-known consequence of an ozone reduction is an increase of ultraviolet light at the ground, resulting in percentages of increases in skin cancer typically twice as large as the percentage of ozone reduction.

But a reduction in ozone also has a number of different climatic consequences. For an ozone reduction of 10%, all these changes are estimated to be of the same order of magnitude as the changes due to stratospheric sulphate absorption up to 0.3°C.

The reason that the effect of an ozone reduction on climate change is so complex is that it has strong absorption coefficients in the ultraviolet and the infrared and that even the weak absorption in the visible area is important because of the large amount of solar energy in this region of the spectrum. Thus, a reduction in the ozone would produce a reduction in the absorption of visible sunlight in the atmosphere, letting more light reach the ground. This would produce warming at the ground.

The stratosphere is kept relatively warm mainly because of ozone absorption in the ultraviolet. If ozone is reduced, less ultraviolet is absorbed, and the stratosphere is cooled. Cooling of the stratosphere can have different effects on the climate near the ground.

The first effect is that the stratosphere, particularly the ozone in it, radiates in the infrared. If the stratosphere cools, less is radiated toward the ground. This cools the ground.

But also, if the stratosphere cools, there are some dynamic consequences. Climatic models generally show that high-level cooling is compensated by low-level warming. One argument is that in the long run, infrared radiation emitted by the whole atmosphere must just balance the total radiation from the sun. Thus, when one part of the atmosphere cools, it radiates less into space; for compensation, the rest of the atmosphere must warm, so that the total outgoing radiation can still balance the incoming radiation.

None of these effects can be computed accurately, and they are all of the same order of magnitude. It is unlikely that they all will cancel so that a net cooling or warming is quite possible at the ground, large enough to have important consequences.

Chlorine

When CFMs (freons) are dissociated in the stratosphere, probably the most damaging result is the creation of chlorine gas. Like NO, this gas acts as a catalyst in the destruction of ozone. The manufacturer's estimate is that, in the absence of contrary legislation, CFM production will increase by about 10% annually. According to uncertain estimates, the result would be about a 9% ozone destruction by the year 2000, more than 35% by 2025. Even now, the calculations suggest that about 1% of the ozone has been removed. Since the natural variations of ozone are much larger, this conclusion cannot be confirmed by observations.

Of course, all the estimates are based on the assumption that chlorine is formed in the stratosphere and that there is no important destruction process. So far, free chlorine has not yet been measured in the stratosphere. Such measurements are a matter of highest priority for the 1970s.

If current ideas prove accurate, some curtailment of CFM production is indicated. If the production and use of CFMs are held constant, the depletion of ozone will reach about 5% in 2000 and 14% in 2025. This, too, is unacceptable. But even if the manufacture of CFMs were to stop entirely in 1995, the eventual depletion of ozone would still reach 15% before it gradually recovers. Thus, it seems likely that the use of CFMs will be restricted in the seventies or eighties.

References

1. Bryson, R. A. and Baerreis, D. A., "Possibilities of major climatic modification and their implications: Northwest India, A Case for Study," *Bulletin of the American Meteorological Society,* Vol. 48, No. 3 (1967) pp. 136–142.
2. Cavender, J. H., Kircher, D. S., and Hoffman, A. J., "Nationwide emissions of controllable air pollutants," U.S. Environmental Protection Agency Pub. AP 115, Jan., 1973.

3. Dyer, A. J., "The Effect of Volcanic Eruptions on Global Turbidity and an Attempt to Detect Long-Term Trends Due to Man," *Quarterly Journal of the Royal Meteorological Society,* Vol. 100 (1974), pp. 563–571.

4. Machta, L., "The Role of the Oceans and the Biosphere in the Carbon Dioxide Cycle," Nobel Symposium 20, Gothenburg, Sweden, August 16–20, 1971.

5. Manabe, Syukuro and Wetherald, Richard T., "The Effects of Doubling the CO_2 Concentration on the Climate of a General Circulation Model," *Journal of Atmospheric Science,* Vol. 32, pp. 3–15, 1975.

6. Stanford, J. L. and Davis, J. S., "A Century of Stratospheric Cloud Reports: 1870–1972," *Bulletin of the American Meteorological Society,* Vol. 55, pp. 213–219.

7. Wilson, Carroll, Ed., "Report of the Study of Man's Impact on Climate (SMIC), Cambridge, Mass.: MIT Press, 1971.

3

Urbanization

and Climatic Change

JOSEPH L. GOLDMAN
*International Center for Solution of
Environmental Problems
Houston, Texas*

Climate, by the very nature of the word implies longevity. We attribute a permanence to climate by relating climate as a particular property of a geographic location. But climate, like the geographic location, is in a state of flux, however, at perhaps a much faster rate than the geological changes that cause changes in location. The implication of permanence to climate is therefore erroneous and should be corrected to imply the various temporal scales of change.

Climate changes as a function of the following natural phenomena:

1. The orientation of the earth with respect to the solar system (the tilting axis).
2. The interaction of the earth with the interstellar space (influence of meteor dust).
3. The internal actions of the earth (volcanic dust).
4. The more timely but less familiar sea-air teleconnections that cause moving perturbations.

Natural phenomena, such as these, are sufficiently cyclic (except perhaps for volcanic dust) to have been documented in some form, and we have learned to anticipate them on some regular basis.

There are, however, acyclic phenomena related to climatic change that may cause changes in climate that can be considered permanent when compared on

43

the same scale of some cyclic phenomena. For example, when considering cycles in the time frame of 2.2, 8, 11, or 50 years a climatic change of the order of hundreds or thousands of years may be considered permanent. While urban complexes are cyclic on the order of thousands of years, they can be considered acyclic when considering changes during periods of 100 years or less.

Urbanization effects climatic change principally by changing the natural environment. If we consider the principle ways by which the urban sprawl changes the environment some are immediately apparent. The first apparent changes are from soft, porous ground or vegetation-covered surface to such hard nonporous surfaces as concrete or macadam. Generally, most of the extensive vegetation is removed leaving only park areas that are about one square city block, or in some cases extending for miles in one direction and a few city blocks wide. In some semiarid climatic areas, bodies of water (reservoirs) are made near urban areas where no lakes existed before. These obvious changes in the surface cause changes to local climate that may grow with further urbanization.

The other, perhaps less apparent, ways in which urban sprawl effects climate are illustrated in Figures 1 through 3 and include the following:

1. Changes in reflective-absorptive character of the surface that cause a temperature buildup within the urban area, called the urban heat island, and that feed back on the moisture characteristics of the surface.
2. Changes in the distribution of heat sources or sinks including the contributions from people, concentration of heating and air conditioning units, as well as the heat generation structures. See Figure 1.
3. Changes in air flow caused by making
   a. ground surfaces rough, resulting in more turbulent vertical wind profiles
   b. large obstructions to air flow that result in changes in both horizontal and vertical wind distributions. See Figures 1 and 2.
4. Changes in the constituents of the air and water
   a. increase CO_2 (more people, less plant life)
   b. changes in chemical constituents due to industry, auto, and so forth.
   c. changes in aerosols content, suspended particulates that have feedback to radiation (photochemical, ozone, NO_2, etc.) or diffuse the solar radiation.
   d. changes in oxygen absorbing life that affect the vegetation and future water circulation. See Figure 3.

Temporal as well as spatial variations of precipitation or temperature have been noted in urban complexes of the United States and in other metropolitan areas worldwide. While not yet fully understood, the anthropogenic (caused by human) effect on temporal precipitation and temperature changes can be divided into two parts: the intermittent changes and the cyclic changes. Intermittent

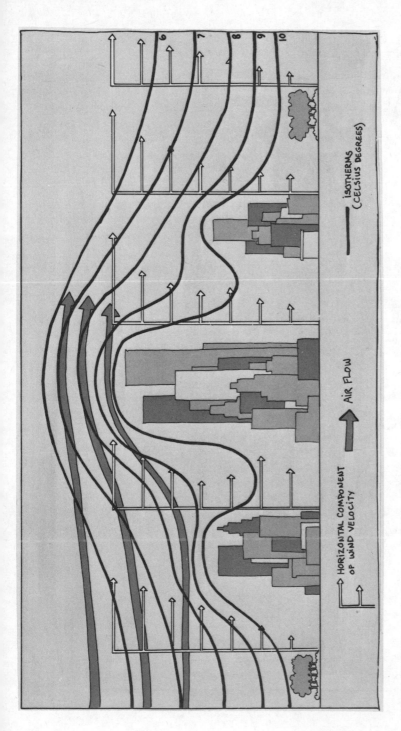

Figure 1. Urbanization affecting airflow and temperature, in vertical section. Effect of obstructing buildings on the profiles of horizontal wind component at various locations between groups of structures. Porosity of buildings may result in the low level peaks of the second and third profiles downwind. Vertically downward winds on the wake side of individual buildings (not sketched) also occurs. Heat retained in concrete buildings and given off together with other heat through heating and air conditioning causes increased air temperature in the proximity. This heating causes isotherms (lines of equal temperature) to rise from the countryside and bulge over building clusters to form the "urban heat island," making air less stable above and more stable within the h at island. Air flow above obstructions generally rises over the urban area due to contributions of the heat and air flow from below.

45

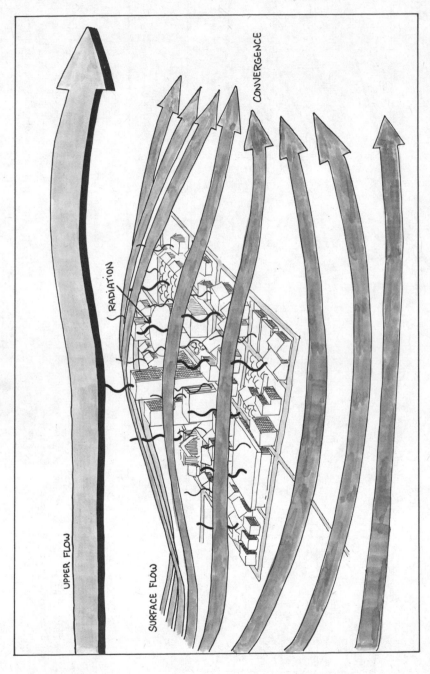

Figure 2. Urbanization affecting airflow. Radiation from characteristically urban surfaces, which expose very little area of natural ground surface or vegetation, causes the resulting urban heat island to act as a mound barrier to surface air flow. The surface air, guided by the upper air, flows around and over the barrier and converges on the downwind side. When the obstructing effect is sufficient, the perturbed surface flow is transported aloft, causing a change in the upper flow downwind of the urban area.

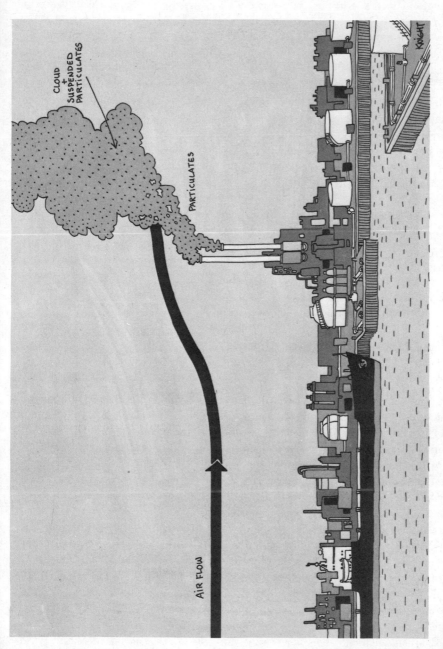

Figure 3. Urbanization effects from high population density industrial sources. Particulates emitted as waste from industrial processes may act as condensation nuclei for atmospheric water vapor. Water vapor condenses on the particles releasing condensation heat that gives rise to cloud convection and supports further cloud development leading to precipitation. The air flow carries the affected cloud within it, giving rise to often observed cloud enhancement and precipitation downwind of industrial sites.

changes may be due to the effect of growing obstruction to the natural flow that permanently disturbs the flow until the obstruction decays. Short-term oscillatory changes may be due to direct, individual effects of altering natural flow by changing the homogeneity of the ground, while longer term oscillatory effects could result from the integrated effects of successive alterations of the flow by combinations of various inhomogeneities. Short-term oscillatory changes can occur by increasing the turbulent wind currents through decreasing the surface homogeneity over which the winds move.

Increasing the difference in ground surfaces—high radiation-reflection and low precipitation-absorptive concrete adjacent to low radiation-reflection and high precipitation-absorptive plowed fields—on the scale of a metropolitan area (e.g., 50 mi in diameter) over many areas spaced at a finite distance apart can cause both decay and amplification of the turbulent convection patterns downwind. These turbulent effects cause changes in the temperature and precipitation patterns that interact with large bodies of water, such as lakes and oceans. Because of their different heat capacities, the water bodies act first as an integrator of the shorter period variable weather and then feed back to the winds a longer period change in temperature that contributes to a greater or lesser input of moisture through evaporation from the water surface and a greater or lesser input of heat from the water surface.

The Importance of the Problem

A Consensus

General agreement exists regarding the importance of urbanization to the climatic change problem. Present thoughts, adequately summarized by Landsberg[23], relate the size of the urban complex with the perturbation it causes, thus leading to the conclusions that continued growth of cities into large conurbations can have regional effects on the climate.[27] This conclusion stems directly from the evidence of the few gathered studies generally restricted to one or a combination of a few urban activities that affect the local weather. The listed effects on the atmosphere that engulfs the urban complex and the area immediately adjacent are the result of radiation balance changes, suspended particulates, heat, and moisture emissions.[12] When air pollution concentrations are increased several orders of magnitude, as they are in urban complexes, sunlight duration and solar radiation intensity are reduced, eliminating nearly all of the shortwave and somewhat less of the longwave radiation.[11] Much more evidence exists regarding the stimulation rather than the inhibition of precipitation. Both pollutants and the urban heat island are related to cloud formation and rainfall in the vicinity of urban complexes. Turbulence is increased, with horizontal wind generally slowed and updrafts induced to increase convection and cloud formation, with summer rainfall enhanced and possibly winter snowfall stimulated. Perhaps the most numerous evidence in case studies is that air temperatures are increased with vertical temperature lapse rates stabilized within the heat island and destabilized above the heat island.

Global climate changes as presently revealed in trends indicate no conclusive proof of urban influences, but there is agreement that human's activities may inadvertently modify the climate (Machta and Telegadas.[26])

In Fedorov's[10] summary of Russian modification work, he pointed out urban activities as possible triggering mechanisms for climate modification. His argument is paraphrased as follows: In dealing with natural forces, we cannot expect to overcome the force through frontal obstruction or opposition. With the concentrations of energy that we consider controllable, heretofore shown to be meager, we can best hope to perhaps deflect the natural forces. This is certainly possible in the atmosphere, as it is in essence a physical medium, however complex, that is a delicate balance of large natural forces. The complexity helps to stabilize interaction, prohibiting the amplification of the many perturbations that might cause permanent changes adverse to the overall balance. However, as we have seen the effect of cloud seeding to stimulate convective growth through latent heat release, so we may find other enhancements to cause self-sustaining perturbations. The chain of events that leads to a particular permanent change in the general circulation pattern that can be called climatic change still escapes our knowledge. The numerous potential climatic changes related to human's activities have been considered in great detail.[28]

The effects are magnified, of course, when the activities are concentrated in a high population density urban complex. What is not clear is to what extent the increased density of human's activities concentrated within the confined space and shortened time (due to the increased activity rate) of the urban environment, can cause climatic change immediately or in short or long time periods. Climate modification rather than an abrupt change of the weather, is embodied in a chain of events that feeds upon itself much like a chain reaction. The successful weather modification process that seems to predominate in Russia so far is the modification of the precipitation process through the enhancement of the fusion part of the process, the decidedly smaller energy producing portion of the system in nature (approximately $\frac{1}{7}$ the amount released through the condensation process). This follows the philosophy of relatively small magnitude efforts that humans, more or less collectively, can perform through aglomeration, causing large scale effects through chain reactions initiated by small scale concentrated trigger mechanisms.

Urgency of the Climatic Change Problem

Climate is the time integrated weather of a given area. To classify the climate of an area, we must specify a horizontal scale and also integrate over a given area. It has been shown by many climatologists and geographers that the microclimate* changes as a function of changes in urbanization. Lowry[25] has shown this for certain cities; Changnon[3] has implied this for La Porte; Landsberg[22] has discussed this in the light of prospects and implications.

This comparison of the more or less permanence of climatic change and perhaps less permanent weather modification can be illustrated by the injection of industrial particulates at a specific rate into the atmosphere that contributes to enhancing rainfall downwind of a moisture source. To render this apparent weather change less effective, all that might be necessary is to change the rate of emission of the particulates (hygroscopic or otherwise) with the wind speed; allowing for a "safe" level of particulate concentration. This could be done conceivably by automatic regulation of the emission rate with the wind speed. When the wind speed is low, scrubbers, precipitators, or other cleansing devices are put into operation. When wind speeds are high the cleansing devices are operating less or are perhaps turned completely off. When we consider, on the other hand, the more or less permanent effects of tall buildings obstructing airflow, concrete surfaces reflecting radiation, the lack of vegetation to support evapotranspiration, balance of CO_2, O_2, and recently O_3—all contributing to a change in rainfall distribution extending from the urban to rural areas—the solutions are no longer merely adjusting the emission rate. Massive changes in surface characteristics, including topography, may be required in addition to the adjustment of emission rates downwind of a moisture source.

The consequences of adverse climatic changes require that we learn to anticipate the changes early, before they are firmly established. We should actively pursue understanding the various causes of climatic change and their complex interaction with feedback mechanisms that make the entire system highly nonlinear. We cannot hope to understand all cause-effect mechanisms, since we must first understand the natural climatic variation, which continues to escape our attempts at successful numerical simulation.[21] Therefore, early indicators of climatic change, some at the earliest possible stage of appearance, should be isolated, classified, then related to measured urbanization for use in extrapolating further climatic change with continued urbanization.

Urbanization Affecting the Environment

Pollution of the environment has been the battle cry of the concerned public, a cry that has recently been losing some intensity because of the energy crisis.

Having received the most publicity, air and water pollution have become the bywords of environmentalists in urban areas. Air particulates and gases have

* The climate of a location usually determined by the temporal records of measurements at a single point representing that location.

caused great concern among the public mainly because of their apparent visibility and in some cases, noxious odor. In certain locales water pollution has also become visible and apparent to other of the senses such as taste and smell. These classically popular effects are treated adequately by Altshuller.[1]

Rainfall Changes

The change in rainfall is an effect of urbanization on the environment that has become recently apparent. Anomalous changes of rainfall that were nebulously attributed to variations in the natural climatic cycle are now being considered for study related to various anthropogenic effects. Increased convective activity that causes the change in urban rainfall results from one or a combination of anthropogenic effects in the collective form of urban characteristics; these change the environment from its natural undisturbed state. The roughened surfaces and increased input of hygroscopic and other condensation nuclei usually result from industrial and other high population density sources. In addition to the heat concentrated in high population density areas, sources of particulates are the result of a large number of people disturbing a small area at a given time, kicking up dust or causing other nuclei to be suspended in the air for condensation.

The role of particulates in the process of condensation is complex—since some nuclei enhance condensation more than others. There are many other complications both in the form of the nuclei and in the amount available in the air at any given time for condensation. Furthermore, salts are hygroscopic, but some salts are more so than others. If dust (silicone dioxide, or some other compound) is suspended in the air, it can act to disperse moisture concentrations before they can become rainfall. Even large concentrations of salts that are hygroscopic nuclei can disperse rainfall by their own overabundance in the atmosphere. This condition results when the available moisture that condenses on the abundant nuclei cannot grow to rainfall size before it is dispersed by the wind. This example of the balance of moisture and nuclei required for rainfall enhancement also serves to remind us that too much of a good thing may be harmful.*

Rainfall Change—Verification Problems

Apparently overlooked until recently is the urbanization effect of changing radiation surfaces and changing roughness on resulting rainfall distributions. Because of a number of data collection and archiving problems, the spatial distribution of rainfall has not been readily apparent using the standard perusal

---

* The rules of moderation seem to extend across the chasm between natural and anthropogenic effects.

procedures. Mainly because of concentration on "class A" weather service station records, changes in the spatial rainfall distribution have not been as apparent as, for example, the changes in thunderstorm frequency.

The detailed data from substations situated around the metropolitan area, which usually contribute to revealing the changing spatial pattern, are not readily available to standard analyses that require only a few representative stations. As with most careful work, effort must be exerted to obtain all available data for a particular area. "All available" requires consulting sources other than the standard climatic data sources and having the incentive to inquire after so-called unauthorized or unapproved local data. The classification *unauthorized* or *unapproved* can sometimes represent National Weather Service stations in a hydrologic study undertaken by the U.S. Geological Survey or U.S. Geological Survey stations in a rainfall study undertaken by the National Weather Service.

With the Houston rainfall distributions, where analysis is restricted to either National Weather Service or U.S. Geological Survey stations, many of the details in the pattern that reveal change are not apparent. A careful study of the Houston thunderstorm statistics[17] indicated the effects of urbanization on climatic change (in this case, the frequency of storms) to be inconclusive.

The particular inconclusiveness of the Houston thunderstorm statistics may have been caused by the change in location of the representative station from the central city to the new airport located over 20 miles from the downtown area, or due to the distribution of the unfortunately few (about 3) thunderstorm reporting stations in the metropolitan area. However, this serves as an example of how the choice of a convenient index for investigating climatic change may conceal climatic change for a particular locale while being both a very effective and convenient designator of climatic change in other urban areas.

Evidence of increased thunderstorm frequency, while implying an increase in showery rainfall, does not necessarily imply an increase in total amount of rainfall. In certain urban areas increased thunderstorm frequency may be consistent with overall increased rainfall, or, on the other hand, increased showery rainfall at the expense of continuous rainfall may result in an overall equal total rainfall or perhaps even reduced total rainfall. The rainfall variety in urban areas is important to considerations of flow rate and erosion that affect flooding, sewer systems, construction, and public health through a myriad of pest control problems.

However, the question of increased versus reduced total rainfall available may effect the very fundamental existence of the urban area, since urban areas usually require larger amounts of fresh water than rural areas. While some highly agricultural rural areas might require more total water, usually this requirement can be satisfied with recycled urban water or perhaps even brackish or other water of sufficient purity for particular agricultural purposes. The only way to determine from rainfall measurements whether the entire rainfall amount has changed or the rainfall has varied in type only is to measure the areal distribution and integrate and compare it with previous areal integrations of the rainfall pattern. This, of course, requires data from a number of stations in the metropolitan area. The more stations in the area, the more detail that can be retained in the analysis, leading to a better representation of the total that has fallen over the area.

Another way to estimate the total rainfall is through measuring runoff in drainage basins. The advantage of this method is that rain that has fallen between collection points can be measured as a contribution to the total water in the particular drainage basin. To estimate rainfall through measures of runoff requires hydrological data including detailed knowledge of the *unit hydrograph* for each drainage basin, one of the ever present *variable constants*. Changes in rainfall type in a drainage basin usually results in changing the unit hydrograph for that basin by changing the drainage characteristics through changes in previously measured erosion characteristics. Other changes in the drainage basin area such as vegetation amount, type, and distribution, time-history of rainfall, evapotranspiration, and wind velocity, also effect the drainage characteristics.[46] Obviously, the hydrologic approach alone, does not simplify the problem of determining whether more, less, or the same amount of water is falling in or near a given metropolitan area. Perhaps, a more accurate measure of the amount of water falling over an area can be obtained by a combination of hydrologic runoff measurements with rainfall measurements. This will provide a comparison of what has fallen with what is running off the land into streams and may also serve to indicate the change in water requirements of the land.

Mention should be made of weather radar as a tool for measuring available water. Radar detects raindrops of a certain minimum size as they are suspended in the air. What makes the radar invaluable to estimating total available water is its capability to scan a volume of the air rapidly without missing any appreciable rain that exists within its range of detection (usually out to 250 nautical miles). There are some drawbacks, however, that must be accounted for in the interpretation of the measurement. Because of a finite beam width, the radar displays a greater area of return than exists, and the relation between target volume and display volume is dependent on the intensity of the return and is not yet completely understood. Since some of the suspended precipitation detected by radar may evaporate before it hits the ground, an accurate quantification of the suspended rain is not necessarily a measure of the amount on the ground.

Using all three techniques of rainfall measurement simultaneously during the morphology of rain events should result in the most accurate measure of rainfall in a metropolitan area. The radar, covering the entire metropolitan area would measure the rain before it falls to the ground and is captured by a sampling network of rain gauges. The rain gauge network will help determine what was lost to evaporation. And the final runoff measurements will determine the amount absorbed by the land and the remaining water available for use by the urban area. Some networks are in operation in certain areas of the United States, England, Australia, and, more recently, one is being instituted in Africa. However, most are operating over vast agricultural areas and generally do not extend over large urban areas. If these networks can be maintained and extended to include the urban areas, more definitive data on urban effects will be revealed, and perhaps quantitative relations will be established on precipitation variation due to urbanization.

We May be Overlooking the Key to Climate Modification

The apparent success of weather modification through the introduction of particulates has naturally influenced the thinking on causes of climate modification. Most of the consideration in climate modification follows the philosophy of the addition or subtraction of atmospheric constituents. For example, one of the principle considerations is with the increase in CO_2 that can effect global temperature.[32] Recently, the effect of the inert gases in sprays[14] on the ozone content of the air* and the continued emphasis on the suspended particulates effecting precipitation are also comparable considerations. Because of the emphasis placed on this philosophy we may be overlooking the key to climate modification.

Fundamentally, climate is based on location latitudinally and on surface type, whether land or water, flat or hilly, vegetated or barren. The general climatic classifications of maritime versus continental and tropical, polar, or arctic applied to air masses are further broken down by geographic and other criteria such as moisture available (in the climatological designations of desert, arid, semiarid, rainforest, etc.). While the latitudinal variation primarily controls the intensity of radiation that effects the temperature, and secondarily contributes to the moisture through circulation, it is the geographic distribution of the land and water at the bottom boundary of the atmosphere that primarily contributes to the control of the moisture distribution and secondarily to the control of temperature through cloudiness. The effect of terrain on the moisture part of climate is manifest in the distribution of tropical deserts and rainforests.

Since the fundamentals of climate are based on the land surface type and configuration, it seems clear that climate variation should be also funda-

---

* An intensive consideration of the implications of changing ozone is given by Haber.[13]

mentally rooted in the change in land surface type and configuration. While the importance of aerosols' effect on the other fundamentals of climate is not, and most certainly should not, be overlooked, perhaps we have been concentrating too much of our precious little effort on the aerosols at the expense of surface type. The various mechanisms by which both fundamental controls affect climate indicate no apparent difference in importance with the possible exception that the direct effects of the land surface type may be more restricted in scale to its source. However, that possible exception may be the result of our limited concentration of effort in exploring the anomalous climatic changes that have not been successfully treated with the aerosol reasoning.

Because of scale considerations, in both time and space, it has been argued that metropolitan areas are too small to cause either lasting effects in time or effects of large enough space scale to change climate measurably. These arguments, when based on conceptualization, correctly point to the size of the urban area and compare it with the size of meteorological systems that are orders of magnitude larger. When based on measurements, such as the recent findings of Dr. Namias and the Norpax Project, the arguments revert to time scales of years (about 8.5 years is a recently discovered teleconnection) and generally portray the cyclic nature of related changes that do not reflect the irreversible changes we expect to be associated with urbanization. The irreversible changes may be concealed within the rate either of urbanization, which also varies with time, or of climatic change, which varies somewhat differently because of inclusion of various natural cycles.

Certainly, all of the climatic change transfer mechanisms from the boundary layer to the free atmosphere are not yet known. Some mechanisms, such as momentum transfer, are being studied and are being seriously considered for verification by measurement programs. Until the time that particular transfer mechanisms can be isolated for the individual phenomena, we must draw from the various possible ways that relatively small perturbations at the earth's surface can affect climate, the time integrated weather that is related to large scale upper air circulation, both locally and downwind.

Hypothetical Case of Urbanization Affecting General Climate

Illustrated in Figures 4a through 4d are examples of how three metropolitan areas along the general track of the weather grow to megalopolis size and affect the low level airflow by causing permanent perturbations of the flow. Suppose this perturbation is transmitted in some manner upward to the upper tropospheric flow and affects the jet stream, the meandering river of air whose character and intensity guides our changing weather systems around the globe. Also, suppose that the spacing of these perturbations causes them to be additive as they are transmitted upward, resulting in a single large perturbation of the

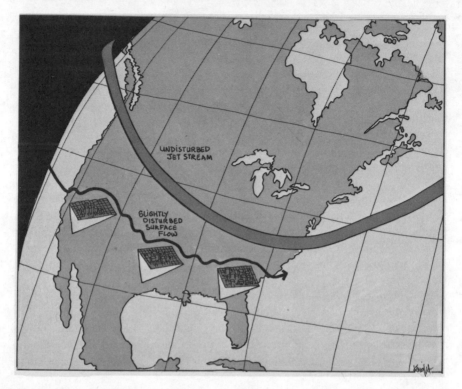

Figure 4. Hypothetical case of urbanization affecting climatic change. (*a*) Urbanization affects surface flow locally; no apparent effect to jet stream.

jet stream. In such a case, when the jet stream would ordinarily be a single wave of a certain amplitude (shown in Figure 4*a*), guiding the number and intensity of storms along its path, it is perturbed to a wave of another amplitude, thus changing the ordinary number, intensity, or path of storms, as in Figure 4*d*.

While the mechanism and processes by which individual weather systems can collectively affect the climate is not yet clearly understood, some speculation in the light of recent reported connections between weather and climate is possible.

It does not seem too unlikely that the increased vertical motion due to the combination of topography and heat emission of metropolitan areas may cause concentrations of vorticity to form in the upper levels of the atmospheric boundary layer. These concentrations may combine as they are carried downwind and vertically by the converging air currents associated with the atmosphere below the Jet Stream. The vorticity can be transmitted vertically

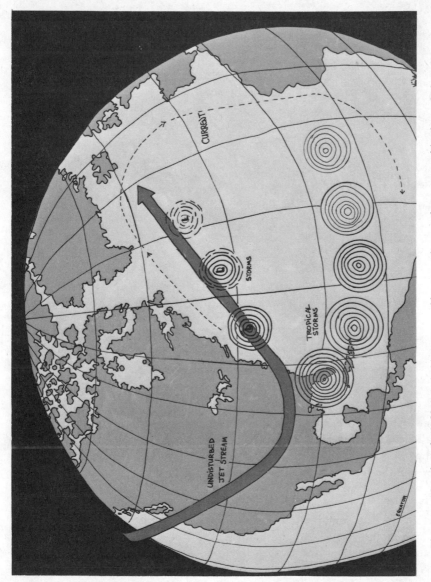

Figure 4. (*Continued*) (*b*) Undisturbed jet stream guides storms in normal manner resulting in tropical storms.

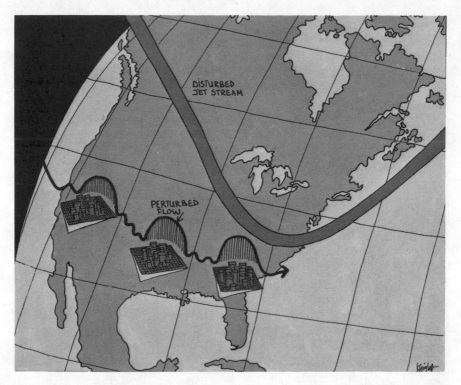

Figure 4. (*Continued*) (*c*) Urbanization perturbs surface flow; in combination with storm moisture causes additive effects at upper level that alter the jet stream.

and added or subtracted to the existing vorticity at the approximate midlevel of the atmosphere (at 500 mbar pressure, or approximately 18,000-ft altitude) from where the present operational numerical prediction techniques guide the further development and motion of weather systems with measurable skill. The adding or subtracting of vorticity affects both the future curvature of the air flow and the horizontal wind shear downwind in the Jet Stream at this level and above.

Thus, the juxtaposition of large urban complexes that cause small scale perturbations to combine into larger scale perturbations can effectively cause changes in the vorticity at the guidance level of surface storms, resulting in changes in the path of storms, their intensity, and in their number.

When more storms occur over a body of water, depending on the temperature of the air relative to the water, they will cause heat to be transferred from water to air or air to water. Expanding somewhat on the findings of Namias[29] and Namias and Born[31] regarding the teleconnection between atmospheric

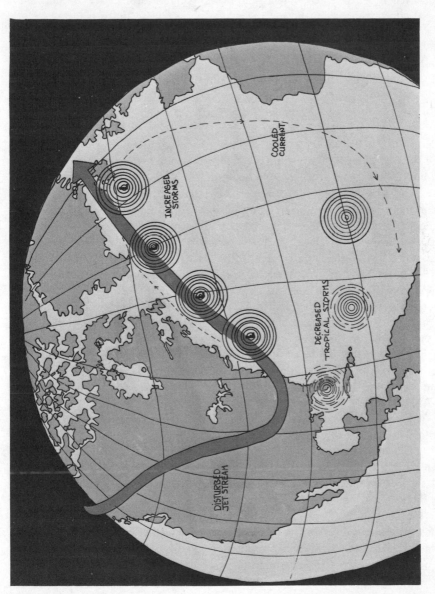

Figure 4. (Continued). (d) Disturbed jet stream increases cooling effect of storms on ocean resulting in decreased tropical storms. With fewer tropical storms, causing less moisture for perturbations, jet stream reverts toward undisturbed shape in (a), beginning another cycle.

systems and sea surface temperature, we can speculate on possible results of the interplay between weather systems and ocean temperature. In areas where the general wind field is from the south, causing surface water to be transported in the same direction (such as our own Gulf Stream), the occasion of north winds would likely be cold relative to the water and cause the extraction of heat from the surface water and thereby the cooling of the surface water. The more frequent and more intense the storms bringing the north winds, the greater cooling effect we would expect.

The recent findings of Namias[30] concerning the role that large bodies of water play to retain integrated temperature effects of successive weather storms leads to this further consideration. The number and intensity of storms over a large body of water in a given time may serve to change the water temperature not only by the transfer of sensible heat, but also by the stirring of water by these same storms, causing local upwelling, and thereby affecting the temperature of more water. Upwelling was found following a hurricane passage over a large body of water.[24] The secondary effect of the more intense storms would be to mix the water vertically, causing the exposure of more water to the air, thus contributing to the effect of a greater water mass.

It is understood that very intense storms such as hurricanes (typhoons in the Central and Eastern Pacific) can spread large amounts of the surface warm water rapidly, causing equivalent amounts of the deeper cold water to rise to the surface. This mixing would expose more water to the air; however, since the rate of cooling is dependent on the air-sea temperature difference, it would most likely be reduced. The physical characteristics of the water, principally its heat capacity, cause it to retain its new temperature as it progresses generally along the path of the prevailing current. This results in cooler water downstream. Since the surface water currents move much more slowly than the air (one-tenth to one-hundredth of the air speed), more water surface is affected by exposure to the cool air downstream.

Air temperature is also modified by continued exposure to the water surface downstream, so that the greatest heat transfer is expected to occur where the cool air first encounters the water surface, at the boundary of land and water (the shore). Evidence of water surfaces modifying the atmospheric wind systems near the boundary of land and water is quite apparent in satellite photos of clouds depicting the sea-breeze front, Bénard convection, row convection, and other organized cloud systems frequently near the Gulf Stream, and less frequently elsewhere. Also, the intensification of low pressure systems (storms) as they progress from land to water near the northern reaches of the Gulf Stream along the East Coast of the United States is apparent in the increased winds that are reported at fixed surface locations. This is especially noticeable along the somewhat meandering boundary between the Gulf Stream and the

Labrador Current. The increased storm intensity is also apparent in the satellite photos of the cloud manifestations.

The effect of surface water on the surface weather system is individually large and short-lived, while the effect of the surface air systems on the water is individually small and in the long-term. Many surface wind systems are required to modify the water temperature of the general current. Also, the motion of the modified surface current is so sluggish, compared with the motion of storms, that continued exposure of the surface water downstream of storms allows for the additive effect of cooling to continue downstream. With water currents progressing at 0.5 knots, storms progressing at 5 to 15 knots and storm winds at 50 knots, it is not difficult to see how increasing the frequency of storms could cause decreasing water temperatures to be carried progressively along the current. This decreasing water temperature would result in modifying the present storms less and less on the near short-term, while retaining less heat for storm intensification downstream in the far short-term. The progressive decrease in the capability of the water to modify air temperature causes the atmospheric wind storms to retain their land characteristics farther along the path over water, thereby resulting in further cooling of the water. However, cooling takes place more slowly.

We can liken the effects of storms on the surface water temperature to rolling a tire up an incline. There is a certain frequency of slaps with our hand of a given strength required to keep it moving up the incline. If we increase the strength of the individual slaps and retain the same frequency, it will go up faster; or if we keep the strength the same and increase the frequency, it will go up faster. If we increase the frequency of storms of the same temperature difference, the water temperature will lower; or if we keep the frequency the same and increase the temperature difference, the water temperature will lower. However, as with most analogies, they only go so far. The difference lies in the reversible nature of the phenomenon. In rolling the tire up the hill we eventually use up the energy we maintained to operate against gravity; the changing storm frequency caused by the perturbing effect of the growing urban areas is an irreversible process, where we are not maintaining energy to operate against nature. Therefore, once the perturbing effect, which is now the cause of changing water temperature, is established, developed, and operating on the air, it remains until another perturbing effect comes along to change it.

To complete the loop required to make the climate change stable, we consider further effects of this hypothetical change in northern latitude storms. A global effect of water temperature cooling by storm frequency and intensity, shown in Figures 4b and d, may be related to the compensating effect of the cooled surface water for tropical storm development. The case of the perturbed jet stream with resultant extensive water cooling can be extended to cause lower surface water temperatures in the return currents to the tropics. When this cooled water reaches the tropical latitudes, it would modify the normally warm water required for maintaining tropical storms. The resulting cooler surface water at the tropics would certainly effect the number and intensity of westward moving tropical storms. Depending on one's point of view, this might be a balancing effect of nature, fewer northward progressing storms resulting in more tropical storms and a greater number of northward progressing storms resulting in fewer tropical storms.

In a certain sense, this may be a natural compensation for increased urbanization, especially in those areas of the southwestern United States that receive a significant amount of their seasonal rainfall from tropical storms. Through decreasing the number of tropical storms, the moisture available from tropical storms is also decreased. Due to the lack of moisture, the urban affected convective activity is inhibited, thereby inhibiting the vertical transport of the urban perturbations to the Jet Stream, and so on through the loop. Since this is a hypothetical case, further consideration of the relationship between the increased instance of extratropical storms and decreased instance of tropical storms may be a curiosity worth investigating.

Examples of Apparent Climatic Change

Preliminary conclusions from the study of 73 years of weather records in a large urban area indicate climatic changes caused by urbanization. Analyses of recent rainfall records in the Houston metropolitan area[6] indicate a change in isohyetal pattern. The normal large scale pattern for the state of Texas shows isohyets oriented north-south, from dry on the west side to wet on the east side, with approximately a 10-in. range across the Houston metropolitan area. In recent years, the pattern has become cellular with a minimum over the central urban complex and large maxima surrounding the metropolitan area mostly on the north and west side, downwind of the urban complex. In 1973 a record rainfall year for 14 metropolitan area stations, records were 250% of normal on the downwind side, while only 110% of normal in the central urban complex. The isohyetal gradient over the metropolitan area is now opposite in sign to the large scale isohyetal pattern, and of comparable magnitude, resulting in approximately 20 in. of additional annual rainfall to the downwind side of the metropolitan area.

With urbanization as the prime cause of the spatial distribution of rainfall over this large metropolitan area, we now inquire as to what urban factors are responsible. We suspect aerosol emissions from both industrial sources and the high population density complex as one possible cause that may affect rainfall by enhanced seeding from below. We also suspect that the increased convection caused by flow obstructions, resulting from tall buildings and the heat-island effect may be responsible. While it is probably a combination of both, the first cause has been receiving much consideration in such places as La Porte, Indiana, through studies by Changnon and collaborators[3] and has been referred to as the "La Porte Anomaly," and in St. Louis, Missouri, through the monumental METROMEX Project, with participation of the Atomic Energy Commission, the Environmental Protection Agency, National Science Foundation, the University of Chicago, and several other universities. Because of cloud seeding, the first cause has been the focal point of scientific popularity as a weather modifier. Preliminary analysis of METROMEX data, revealed after the Houston study results, indicated that the second cause, the mechanical effect of obstructing buildings and heat effect of urban sprawl, may be as significant as the first as a weather modifier and becomes more explicit in the longer term as a climate modifier. The complexity of the causal problem revealed so far from the study of METROMEX data may be somewhat reduced by the comparably simple topography and much less complicated microclimate in the Houston metropolitan area. Its centralized distribution of industrial sources of pollution and its centralized high population density help simplify the pattern of anthropogenic contributions to the urban environment.

Patterns of rainfall excess and deficit adjacent to each other that are manifest as deluge and drought are found worldwide with no particular latitude of the earth apparently free from this intense variation. The tropical latitudes, especially along longitudes of the African desert areas, contain spatial patterns of drought adjacent to deluge that are changing in a cyclic manner. Desert areas or semiarid areas are very sensitive to small changes in rainfall amount that are translated to large changes in percent of total rainfall. The Sahel, the now infamous southern border of the Sahara, is experiencing climatic anomalies that may be related to changing surface conditions (Charney and Stone[5]). Urbanization may be contributing to the anomalous climate of the Sahel by affecting the circulation pattern at other latitudes. In the desert areas, entire nations are dependent on the rainfall during one month, however small that may be in total, to harvest their principle crop.[33] Because of this sensitivity, and the small base for comparing rainfall changes, perhaps the desert areas are not good indicators of climatic change.

In the less sensitive areas of Africa, where larger rainfall amounts are normal, similar climatic anomalies of rainfall excess and deficit have become apparent. The February 1974 rainfall in South Africa (30 degrees latitude south, climatically the same as Houston's) near urban complexes, indicated less-than-normal rainfall within 20 miles of 10-times-normal rainfall. This pattern was noted as developing from a simpler, more homogeneous rainfall pattern related to the topography and to the water source locations of that area. This example of climatic change may be evidence of a long term gradual change that appears intermittent because of the growth of urban complexes. If so, the variation in pattern caused by the urban complexes will eventually stop and may disappear (on the time scale of 100 or more years).

A recently discovered increasing temporal variation of conservative indices from long term normals may also be related to urbanization effects. Temporal variations of rainfall averages from long-term normals have been increasing. The deviation from normal rainfall (abnormal rainfall) has been increasing with time over a period of more than 10 years. In one case, decade averages of rainfall (considered to be quite conservative since they account for the sunspot cycle and biennial cycles), near a large metropolitan area indicate an increasing tendency to be farther, both above and below, from the 30-year normals, the standard period used for established normals. This tendency should be checked in all major metropolitan areas with more than 50 years of data and compared with urban growth to determine whether variable decade rainfall is a signature of urban interference in normal climatic change.

Speculation on Climate Changes

The similarity of changing climate found at Houston and at urban South African areas at the same climatic latitude, approximately 30 degrees north and south, led to further speculation on anthropogenic effects on climate, worldwide. When treated individually each urban complex contribution is small, but each contribution can be transported zonally (west-east) through the capability of the boundary layer winds, and then integrated by the oceans to be delivered farther downwind as a metropolitan scale anomaly in precipitation and temperature that is slowly moving around the earth (at approximately one cycle in 8 years). The 8-year cycle, when added to the known extraterrestrially based cycles such as the variable 11-year sun spot cycle and the biennial oscillation, should result in a better verified rainfall outlook.

The lack of highly industrialized areas in the tropics leads to further considerations that the measured 8-year cycle of climatic change may either be transported meridionally as well as zonally, or may be related to natural effects rather than effects caused by people.

Evidence of modern man's effect on climate is increasing, and we need to quantify the extent of the effect and discover how to accommodate the effect. We see evidence of climatic change that may be due to small disturbances of natural processes that are amplifying. Natural forces are so large compared to man-made forces that it is unrealistic to expect to accommodate man's effect with directly opposing man-made forces. With two massive equal opposing forces, a relatively inconsequential additional force can control the resultant in any direction. So, the direction of the resultant massive natural forces might possibly be changed by the man-made disturbances indicated by our early evidence of urban effects. If this connection of problem with solution is true, we had better learn to direct our urban affected climatic disturbances toward rather than against natural change.

Temporal Variation of Temperature

Recently Dr. Jerome Namias,[30] an eminent research authority on long-range forecasting, related changes in the North Pacific sea surface temperature with warm winters in the eastern part of the United States and cold winters in the west. He explains the relationship as follows:

Since atmosphere-ocean interactions, especially the heat exchange, occur chiefly at the air-sea boundary, sea surface temperature pattern recurrence would seem to play an important role in generating persistently recurrent atmospheric anomalies. For example, fronts and cyclones can be identified in areas of strong anomalous sea surface temperature gradients. Anticyclones are encouraged in certain anomalous areas where the stability in the frictional boundary layer is influenced by the temperature of the underlying surface waters.

Beginning in the winter of 1971–1972, an abrupt change in North Pacific sea surface temperature and atmospheric patterns occured, leading to warm winters in the East and cold winters in the West. These changes were related to changes in sea surface temperature patterns over the North Pacific. The temperature fluctuations are strikingly illustrated in Figure 5, a graph of the mean winter temperatures at representative stations in the eastern part of the United Sates, Nashville, Charleston, and New Orleans.

In Houston, as shown in Figure 6, instead of the winter of 1972 being the warm one, the sharp rise occurred in 1971, followed by a slower rise in 1972. The sharp decline between 1972 and 1973 is significant. The steady increasing trend from 1968 to the high in 1972, which was near the top of the range, resulted in an anticipated sharp readjustment, such as occurred in 1973. Sharp declines following increases had previously occurred in 1951, 1953, 1958 and 1963. The decline in 1973 was accompanied by 3 snowfalls, a record for any single season.

The North Pacific sea temperatures were close to normal during 1972, and therefore, perhaps not a good precursor for the Houston change, however, in contrast, sea surface temperatures in the North Altantic Ocean during 1972, decreased abnormally in April, as illustrated in Figure 7. The temperature decrease described as a cold water eddy, served to enforce the high pressure ridge in the North Atlantic. The enforced high pressure resulted in a blocking

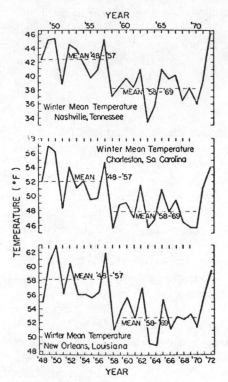

Figure 5. Winter mean temperatures at Nashville, Charleston, and New Orleans from 1947–1948 (labeled 48) to 1971–1972 (labeled 72).[28]

action, which forced low pressure systems (storms) to go northward rather than take the usual northeastward path. In general, "blocking highs" occur during warm trends in the southeast section of the country. Because of the blocking action, fewer storms bring low temperatures to the Gulf of Mexico Coast, and higher temperatures along the Gulf Coast serve to increase precipitation since warm air holds more moisture.

Following the reasoning of Namias and Born[31] that the large bodies of water provide a reservoir for climatic feedback, with time lags of up to two years, perhaps the Gulf of Mexico should be studied for sea surface temperature anomalies. Sea surface temperature correlations are influenced by large scale ocean current systems, and the greatest zonal currents of the North Pacific are regulated by the zonal wind speeds. Anomalies in the oceanic heat reservoir must be considered in the quest for improved long-range weather forecasting. Perhaps the Caribbean Sea and the tropical Atlantic that are governed by the prevailing wind currents can be shown to contribute to the changing temperature of the Gulf Coast area.

Urbanization Effect of a Large Megalopolis

Prompted by local flash floods and short period (3 and 6 hour) record rainfall, a detailed study of the temporal variation of rainfall in Houston was begun in June of 1973. The initial results indicated the need for both further study of the temporal variation of rainfall and an extension of effort to the spatial variation. With continued occurrence of high rainfall during 1973, the temporal and spatial variation studies continued, were documented,[7] and preliminary results summarized by Crooker and Goldman.[8]

These studies led to the formation of an interdisciplinary group to consider the implications of the variation in rainfall to the Houston metropolitan area.

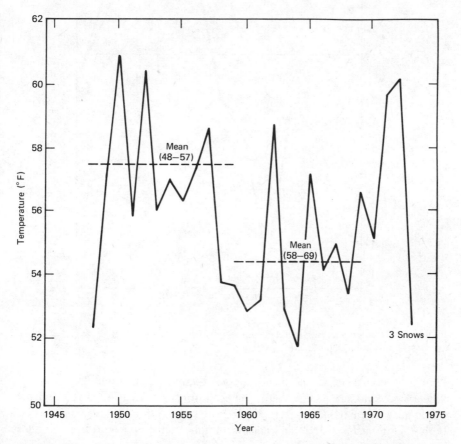

Figure 6. Winter mean temperature—Houston (city), Texas (1948–1973) for comparison with Figure 5.

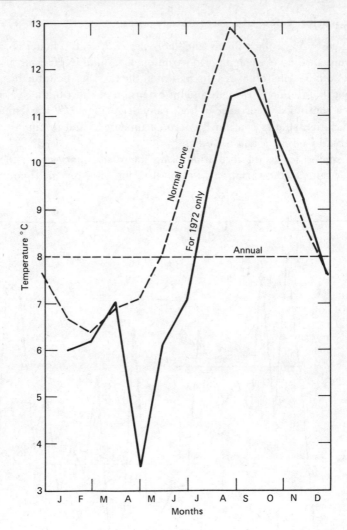

Figure 7. Monthly mean sea surface temperature—1972. Ocean Weather Ship C. Latitude 52°45′N, Longitude 35°30′W. North Atlantic.

The group considered the problems of unanticipated rainfall and locally accelerated land subsidence as they effect future growth of the area and prepared general guidelines for establishing the trends. In an extensive proposal,* both the immediate and far reaching implications to its future growth of

* *Environmental Assessment for Houston, the Future Megalopolis* (available from the Institute for Storm Research, Houston, Texas).

Houston's change in rainfall were outlined and guidelines for implementing verification studies and additional exploratory studies were presented. These studies spanned the broad spectrum of environmental subjects, from behavioral responses to gray days to detailed geological study for drainage and catch basins, all directed toward economic assessment of the possible change in climate.

The interdisciplinary group, composed of a corporate economist (also a behavioral expert), a civil engineer (municipal drainage expert), a civil engineer (marine sediment expert), medical doctor (also a waste utilization expert), and a meteorologist (specializing in mesometeorology), prompted an investigation of Houston's urbanization growth. The initial implications were that the rate of growth was related to the changing rainfall pattern. This relation, examined further with the cooperation of the Houston City Planning office, verified the initial results and was summarized by Crooker and Goldman.[9] In their preliminary study (Crooker and Goldman[8]), the change in climate of the Houston metropolitan area was revealed through a noted change in rainfall gradient. The cause of the rainfall change was attributed to three anthropogenic changes in the surface boundary layer, changes in atmospheric constituents through addition of aerosols, the changing radiation-absorption pattern through nonporous concrete surfaces, and the change in roughness characteristics through the obstructing effects of tall buildings and their heat patterns. The last cause was considered primary in the Houston metropolitan area where homogeneous local topography magnifies the importance of tall airflow-obstructing structures. However, in other major metropolitan areas with variable terrain, the other causes may be more important.

The change in climate of the Houston metropolitan area originally suspected through changes revealed in the temporal variation of rainfall at the official Houston station, was also apparent in the study of other stations in the extensive metropolitan area. Semi-decade averages of rainfall at two stations, one generally downwind of the other, were chosen to illustrate the details of changing climate as a function of increased rate of urbanization.

To establish a base from which to compute comparative changes, the Texas Upper Coast, a homogeneous area for which statistics are prepared routinely and containing the two stations, was chosen.

The state of Texas is subdivided into the homogeneous areas shown in Figure 8. Differences in latitude, longitude, and elevation of the ground above mean sea level are shown in Table 1, below.

The topography of the Houston metropolitan area, shown in Figure 9, illustrates the terrain with major airflow disturbances and the location of the source of pollutants or aerosols near the industrialized section of the Ship Channel. Cypress is located 25 mi northwest of the business area of Houston.

The Cypress record began in 1943, while the official record for Houston

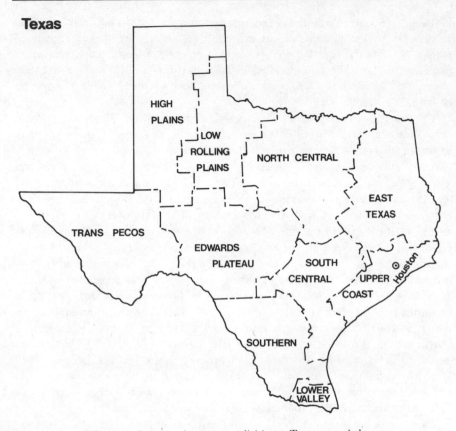

Figure 8. Relation of upper coast division to Texas as a whole.

(City) extends back to 1882. Double-mass analysis of the Cypress station permitted the record extension to 1941 for comparison with the longer Houston record. Data normals (1941 to 1970) for the Texas Upper Coast were used as base values for the comparison. Since 1970 precipitation has been noticeably progressively heavier in northwest Harris County, resulting in Cypress receiving 92.62 in. in 1973 while only 59.53 in. fell in downtown Houston.

Table 1.   Index

Station No.	Name	Latitude	Longitude	Elevation (ft)
1	Houston-City	29–46	95–22	41
2	Cypress 1 SW	29–58	95–43	150

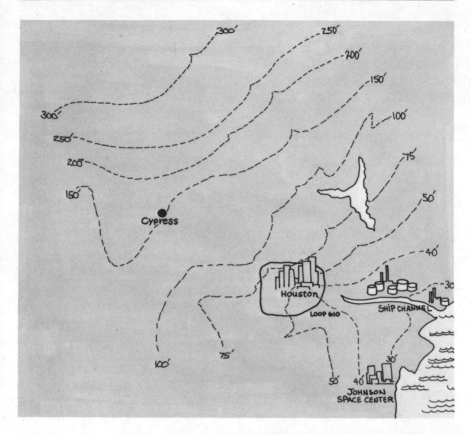

Figure 9. Topography of Houston, Texas. Elevation contours, in feet above sea level, illustrate the gently sloping terrain, whose greatest two disturbances are from high rise buildings at the Johnson Manned Spacecraft Center (mostly below 250 ft) and from within the central metropolitan area identified by Loop 610 (mostly below 800 ft). Cypress is a rainfall station downwind of the Houston central City station, used to illustrate the effect of urbanization on the metropolitan rainfall pattern.

The ratio of Cypress and Houston rainfall to the Upper Coast rainfall, shown in Figure 10 for the period 1941 to 1973, indicates that from 1956 to 1960 the rainfall amount was similar for both stations. From 1961 to 1965 Cypress was 108% of the value for the Upper Coast, while Houston's ratio declined from 105 to 96%. In the 3-yr period 1971 to 1973, Cypress was 120% of the Upper Coast normal, while Houston was only 90%.

In 1973 the ratios for Cypress and Houston were 135 and 87%, respectively. Such a shift in pattern, with the same base line, suggests that urban changes must be the major cause of the change in rainfall regime.

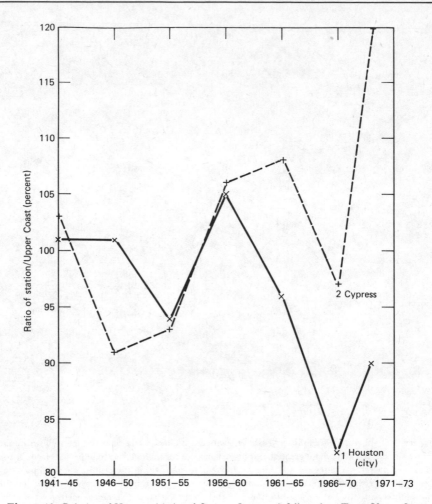

Figure 10. Relation of Houston (city) and Cypress 5-year rainfall totals to Texas Upper Coast.

Metropolitan Area Boundary Layer Changes.   As urbanization progressed, heavy industry concentrated around the Houston Ship Channel where transportation facilities could expedite their products to world markets. Generally, light winds occur from the east and southeast in the early morning hours and during transitions after cold frontal passages and contribute to periods of air stagnation; with daytime heating, the morning temperature inversion is raised and winds become southerly. The resultant wind pattern carries pollutants to the northwest section of Harris County, where they remain suspended until the next cold front causes wind and moisture buildups and eventual rainfall.

Addition of Aerosols. The ship channel area includes a complex of petrochemical and other chemical manufacturing industries, which contribute emissions of particulate matter, sulfur and nitrogen compounds as well as hydrocarbons. Whenever the relative humidity is high, as in the early morning hours ahead of sunrise, the moisture in the atmosphere combines with by-products to form mist-like plumes. These aerosols, along with emitted particulates, form condensation nuclei, which increase precipitation in a manner similar to artificial cloud seeding.

The wind circulation pattern preceding the approach of colder polar air from the northwest produces an area of convergence in the trough of low pressure. Many of the polar fronts become quasi-stationary in the Texas Upper Coast, contributing to a great accumulation of potential moisture in the form of precipitable water, when given the appropriate instability and forced lifting necessary to produce precipitation.

Prior to the accumulation of smoke from the industrialized section, the occurrence of morning fog in this area was usually dissipated within a few hours following sunrise. Since the growth of urbanization, the combination of smoke with fog has extended the time necessary for dissipation to noon or early afternoon. The density of the combined smoke and fog restricts the effects of incoming radiation so that temperature inversions last longer over the eastern portions of the city.

Radiation-Absorption Pattern by Nonporous Surfaces. As a result of a cooperative effort between the City of Houston and the U.S. Geological Survey, Johnson and Sayre[19] concluded that complete urbanization increases the magnitude of a 2-year flood ninefold, and increases the magnitude of a 50-year flood fivefold. These effects are caused by changing rural areas with pasture land to concreted suburban shopping centers, which increases radiation and decreases absorption. Porous soil surfaces are replaced with nonporous concrete surfaces. The changes in surface porosity made by complete urbanization are attended by the increased runoff mentioned; at the same time, when the decrease in surface porosity resulting from complete urbanization is combined with the northwestward shifting of the maximum rainfall pattern, both the magnitude of the runoff may be more than doubled and the drainage area for excessive rates increased significantly.

Change in Roughness Characteristics. The local topography of the Houston metropolitan area is rather homogeneous and consists of very flat coastal plains, which extend approximately 100 mi. inland from the Gulf of Mexico. Urbanization, especially since 1960, has included high rise construction in the central city business district and in outlying business centers. Tall buildings obstruct airflow and interact with the flow, forming waves and eddies on the downwind side. Venturi effects (increased speeds) occur as surface air is confined to blow between buildings, and the Monroe Effect, as air is forced downward to the street level, causing high speed reflex reactions. In addition, these obstructions, with their roof top exhausts of air-conditioning heat of condensation, cause air to rise, which eventually cools, condenses its moisture and forms convective clouds downwind.

Urbanization Determined from Major Office Area.   Using a study conducted by the Houston City Planning department (Jones[20]), changes in urbanization were estimated from data and projections of floor space and land areas of major high rise office centers for 1990. Major office areas, identified in Figure 11, were used as an index of urbanization change.

Urban Development.   Prior to 1960 most of the high rise development was confined to the central business district. In 1960 the central business district (CBD) had approximately 10 million sq ft of office space available compared to about 3.5 million sq ft available outside of the CBD. In 1970 the CBD had 15.7 million sq ft compared to 13.7 million sq ft outside of the CBD. This means that in one decade (1961 to 1970), the CBD increased only 58%, while outside the CBD, major office space nearly quadrupled. The major extension of high rise structures were southwest and west of the central business district, particularly including the medical center complex on the southwest and along Westheimer to 610 West Loop. After 610 West Loop was in operation, large office structures were started on a north-south orientation along West Loop.

Urbanization Data and Projections to 1990.   From data obtained through the Houston City Planning Department, we have compared development in the central business district, the area outside the CBD, and the West Loop. These areas have been projected by the City of Houston to the year 1990, and are illustrated in Figure 12. The rate of increase in the area outside the CBD between 1960 and 1970 can be assessed by comparing the slope of curve *B* with curve *A*.

Projections for the period 1970 to 1980 indicate an increase in the urban growth in the CBD. However, from 1980 to 1990 the trend is expected to reverse again. Total office areas by 1990 have been projected to 140 million sq ft, with only 47 million sq ft in the CBD.

Unfortunately, the areal distribution of reliable, quantitative data was not available for the major growth period, from 1940 to 1960. However, indications from the growth in high rise office space during the 1961 to 1973 period imply a direct connection to the widening difference in precipitation between the Houston City station and the downwind station, Cypress. Related to the large increase in high rise construction west of the CBD (quadrupling during the initial development, 1961 to 1970, with a projected annual increase of 24% for the decade 1971 to 1980), the downwind station northwest of the CBD

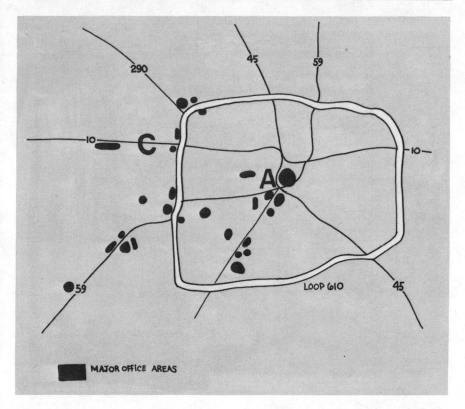

Figure 11. Location of major office areas in the Houston metropolitan area. (Courtesy R. H. Jones, Director, Houston City Planning). Within Loop 610, area A designates central business district where both the tallest and greatest number of buildings are located. Projected growth of the high population density, heat emitting, and wind obstructing structures show a much greater percent increase in Area C, west of the central business district.

increased to 120% of the average rainfall for the extensive Texas Upper Coast division, while the Houston City station, located in the CBD, decreased to 90% of the Upper Coast average. Furthermore, the separation between the rainfall percentages, as indicated in the semi-decade averages, increased gradually during the decade 1961 to 1970, while the high rise construction was just getting underway and the west was becoming significant, then accelerated during the first third of the 1971 to 1980 decade, reflecting the expected acceleration in high rise construction which has been verified at least qualitatively through 1973.

The implications are that increased high rise construction at an average rate of 6.5% annually in the CBD (area A of Figure 11) and 24% annually west of the CBD (area C of Figure 11) will continue to cause rainfall maxima to be

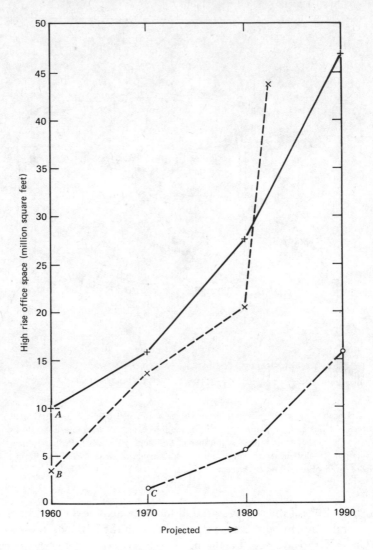

Figure 12. Urban development measured by major office space in the Houston metropolitan area. (*A*) central business district, (CBD), (*B*) outside CBD, and (*C*) West Loop.

shifted northwestward (downwind), further changing the rainfall gradient across the metropolitan area that was originally dry on the west and wet on the east. The extension of the high rise barriers westward may increase effective rainfall in drainage areas over 100%.[18] If this is beyond the capacity for which they were designed (requiring a safety factor of 2 or less) flooding can result leading to the many pest control and other health related problems.

Both the past evidence of change and the implications related to changing urbanization make Houston a good example of a changing megalopolis downwind of a moisture source. The apparent clarity with which Houston has displayed its changing rainfall pattern as a function of urbanization makes it a likely candidate for further study.

Case in Point for a Rapidly Growing Urban Area

The less popular effect of urbanization on climatic change, the effect of obstructing airflow is shown categorically in the case of Houston, a relatively flat area whose major topography is made up of tall buildings. In Figure 9 the contours above sea level indicate that the slope of the land over the Houston metropolitan area, shown approximately enclosed by the county boundary, is indeed shallow, ranging from about 30 to 100 ft in about 60 mi. The slope of the ground surface increases somewhat to the northwest to about 30 ft per 2 mi, still a rather shallow slope. The inner-city area is contained within the loop (highway indicated by I-610), and there are several groups of tall buildings with the principle group located in the downtown area, approximately at the center of the enclosed area. The general locations of other major areas of tall buildings and industrial sites are indicated.

If the more popular cause of precipitation variation, the injection of pollutants that can enhance or inhibit rainfall, were predominating, the rainfall excess would be restricted to an area downwind of the ship channel from where a high concentration of chemical and industrial pollutants emanate. That downwind area would be north of the city since the prevailing winds are generally from the southeast or south-southeast. However, as has been shown (Crooker and Goldman[8]), areas west of the metropolitan north section have been rather significantly affected in rainfall; some areas to the southwest of the industrial center have also been affected to the degree that other influences must be operating to control the rainfall distribution.

Recent research results indicate that Houston's apparent urbanization effect on rainfall may be characteristic of urban areas in general. Similar rainfall enhancement by the obstruction effect has been found recently in Columbia, Maryland (Harnack and Landsberg[15]). The recent results of the Metromex Project[4] also agree that the obstruction effect may be more important than the aerosol effect on rainfall distribution.

Temporal Variation

Decade averages of precipitation, which take account of the effect of the approximate 11-year sun spot cycle* on the temporal distribution, in Houston

---

* The sun spot cycle and its possible relationship to meteorological phenomena may be found in Roberts, et al.[39]

from 1901 to 1970 indicate a sharp rise for one decade followed by a slower decrease for two decades, as seen in Figures 13 and 14. The decade 1961 to 1970 represents the end of a two decade decline.

During the first three years of the present decade (1971 to 1980), precipitation in the Houston area had been increasing, each year wetter than the previous.

Rainfall in 1973 was particularly heavy in April, June, September, and October. The Houston official annual rainfall total, measured at the Intercontinental Airport, north of the city, was 70.16 in. in 1973, which is 153% of the normal. Significantly high percentages occurred in four months as shown in the following tabulation.

1973	Percent of (1931–1960) Normal
April	124
June	380
September	232
October	250

Spatial Variation

The normal large scale areal rainfall distribution for the state of Texas,[34] illustrated in Figure 15, is generally a linear pattern of dry on the west and moist on the east, with about a 10-inch range across Harris County, the Houston metropolitan area. Focusing on Harris County, containing the Houston metropolitan area, the updated normal rainfall pattern as shown in Figure 16 retains the general linear pattern of the large scale statewide distribution, with perhaps a somewhat greater change from west to east across the metropolitan area. The minimum in rainfall indicated by the closed 45-in. isohyet in the center of the distribution is a result of the rainfall at the central city station (HOUC). Recently analyzed patterns of rainfall determined from combined data from the urban runoff study conducted by the City of Houston and the U.S. Geological Survey with the National Weather Service data indicate a change from the normal linear pattern to a cellular pattern.

On an annual basis, precipitation in the Houston metropolitan area varies considerably in space, and the variation is more cellular than linear. The stations used in the detailed analysis of the spatial rainfall distribution and their distribution around the densely populated area are shown in Figure 17.

At the end of November 1973 Independence Heights, a substation approximately 7 mi from the City office, had received more than 14 in. above the city record. The variation becomes still larger when extended to include

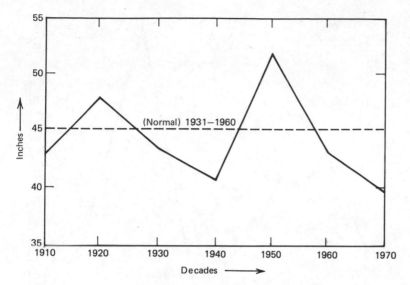

Figure 13. Decade precipitation for Houston. Graph of 10-year averages during the period of record, 1901 to 1970. Note the repeated cycle over 30-year increments.

Angleton (just off the map in Figure 17, to the southwest), Cypress, and Spring Branch. In 1973 Angleton received 100.21 in., Cypress 92.62, and Spring Branch 71.45; each of these a new record.

Spatial analyses of rainfall containing the cellular pattern indicates that the central part of the city of Houston is not representative of metropolitan area rainfall. For example, in the rainfall analysis of the hydrological water year, ending September 30, 1970,[45] there are areas in the southeast, western, and northern portions with annual rainfall equal to or exceeding 50 in., while the center of Houston received only 32 in.

The rainfall pattern for the calendar year 1970, Figure 18, contains a similar pattern to the water year, with the general increase from 40 in. in northwest Harris County to 55 in. in the east. The cellular pattern seems to dominate the spatial distribution, and with the annual wind from the southeast, there is a cell of 55 in. downwind from the 40 in. in the downtown area. Figures 18 through 28 show annual precipitations for the Houston metropolitan area. In each figure X locates station data available during the analysis; isohyets are in inches; and wind speed is in miles per hour.

In 1971, shown in Figure 19, a below normal year, wind was still from the southeast and the center of the city was drier than the 3 cells reporting over 45 in. in the western half of the county.

Conditions in 1972 (Figure 20) with wind from the east-southeast, shows a

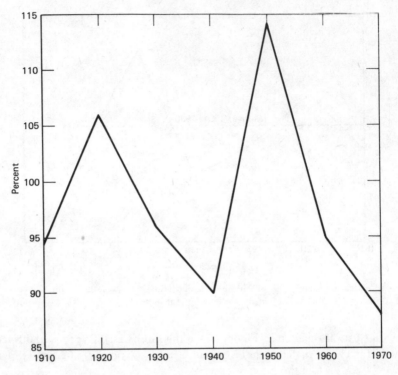

Figure 14. Decade precipitation trend for Houston (1901 to 1970). Percent of decade averages to 30-year normal. Note that changes by decades between 1901 to 1970, in percent of the 1931 to 1960 normal, show ranges from 88 to 114%. The low value of 88% for the decade ending in 1970, implies a rising trend is in the immediate future.

marked increase in rainfall and a more complex cellular pattern. An area of 60 in. prevails northwest of the downtown area, and two dry cells of 40 in. prevail in the West and near Tomball.

Rainfall in 1973, shown in Figure 21, indicates not only an overall increase in amounts, but also a more complex cellular pattern. Wind for the year was from the southeast as for most years. The downtown area had only 59.53 in. compared to 74 in. just about 7 mi. northwest at Houston-Independence Heights. In northwest Harris County, Cypress received 92.62 in., a new record, which exceeded the previous record set in 1960 by more than 15 in. A drier area did occur, however, in southwest Harris County with 59 in. recorded at Houston-Addicks.

In the four successive years, 1970 to 1973, the cellular pattern of a minimum in the downtown area with surrounding cells of maxima prevailed. This pattern, apparently a permanent feature of the Houston metropolitan rainfall, has the northwestern area, downwind of the downtown urban complex usually covered by more rainfall than the central area. The results of these initial findings led to a consideration of the changing "normal" pattern and to the question of when and how the pattern changed.

Normal Rainfall. In recent years, normal periods have been standardly computed from 30-year periods of record for comparison over wide areas. In the earlier 1940s, normals were computed from the entire period of record, but this practice created bias when comparing long records with short records, for example, in comparing New England with more than 100 years of records with stations in Texas or western states with 10 years of records.

Until data were processed through 1970, normals were based upon a stand-

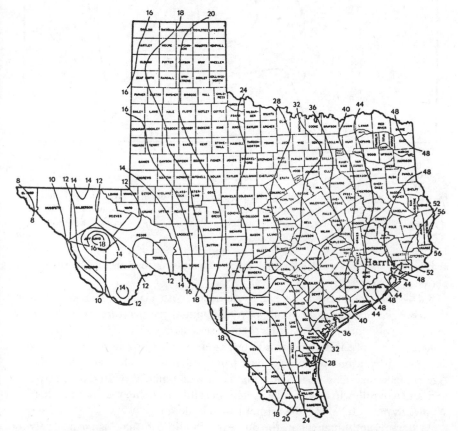

Figure 15. Normal annual precipitation in inches. Based on the period record, 1931 to 1960.[38] The precipitation distribution from west to east across Harris County, containing the Houston metropolitan area, is dry to wet.

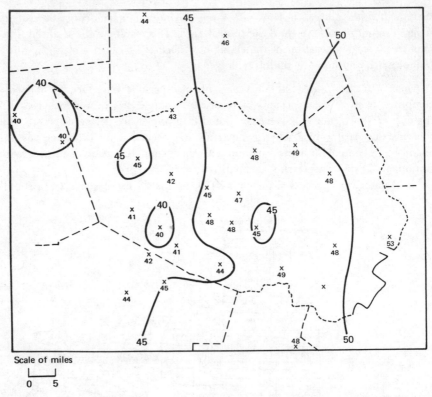

Scale of miles

0   5

Figure 16. Average annual precipitation for the Houston metropolitan area, Harris County, and adjoining counties. Normals computed from data available by National Weather Service for period of record 1941 to 1970.[33] Isohyets drawn for each 5 in. Prevailing wind velocity SSE 10.8 mph.

ard period of 1931 to 1960. The latest normals in use in 1974 are for the period 1941 to 1970.

Such shorts periods should not be confused with geological epochs such as the Ice Age. Bryson[2] in his study of mean annual temperature in Iceland over the past millennium, compared growing season, sunshine, July temperature, snowfall, and precipitation for 13 thousand years. Long term relations were determined from implications derived from pollen data. He concluded that Ice Age climates may end in a century or two, while glacial and oceanic response and a new equilibrium may take millennia. Holocene climatic changes that are smaller in magnitude, may be accomplished in decades.

As Bryson aptly points out, the 30-year climatic "normal" is normal only by definition, "While some causal parameters, such as earth-sun geometry, change slowly, others such as volcano-induced turbidity, may change rapidly and sporadically."

A principle result of the Houston rainfall pattern change with time indicates that the heavy rainfall pattern is being displaced to the northwest of the center of the city, which coincides with the usual wind flow from the southeast. To illustrate the change in spatial pattern of annual rainfall throughout the record period, a number of years have been selected at random to represent the decades prior to the present decade. The years selected are 1908, 1913, 1929, 1935, 1946, 1958, and 1966.

Detailed Rainfall Analysis

Because of the limited reports available before 1940, surrounding stations in other countries were used to help determine the location of isohyetal lines (lines of equal rainfall) as they crossed Harris and adjoining counties.

In 1908 a comparatively heavy rainfall center, shown in Figure 22, continues

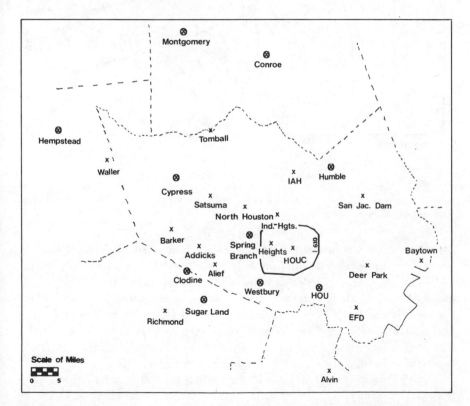

Figure 17. Location of Houston metropolitan network stations. Loop 610 encloses the population dense portion of the metropolitan area. Circled stations indicate where new records were established in 1973.

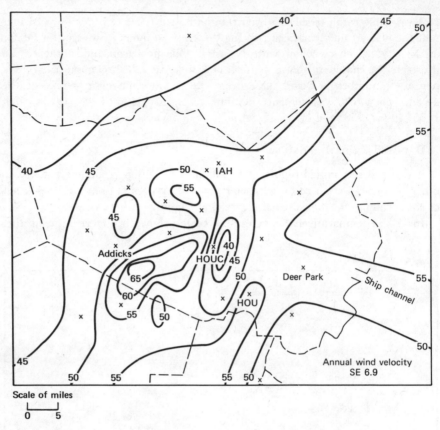

Figure 18. Annual precipitation in 1970 for Houston metropolitan area, Harris County, and adjoining counties. X locates station data available during this analysis. Isohyets are in inches; wind speed in miles per hour.

from Fort Bend County southwestward to 56 in. at Pierce near the Colorado River. The heavy rainfall pattern in Liberty County extends eastward to 60 in. at Beaumont. One obvious conclusion of this pattern is that there is an inadequate network to show distribution in Harris County. Since 1908 was before the establishment of a Weather Bureau Office in Houston, the wind record is not available. While the heavier rainfall pattern in Fort Bend County is the outstanding feature of this map, a substation in Houston, location unknown, reported 49.11 in. in 1908, which was down from 62.51 in. in 1907.

The 1913 map (Figure 23) has above normal rainfall in the northeast portion of the county. The increase in Waller County continues to 55 in. at Somerville, then eastward, 66 in. occurred at Beaumont and Port Arthur. In 1913, years before flood control storage was made effective, the Brazos River flooded over an extensive area. The prevailing wind from the northeast, an unusual prevailing wind direction, caused the moist tongue and westward curve of the 55-in. isohyet. The annual total rainfall downtown was 54.50 in. compared with the normal of about 45 in., while, in general, rainfall was increasing from 45 in. in 1912 to 61 in. in 1914.

The 1929 map (Figure 24) locates heavy rain in Montgomery County, with a maximum pattern northeastward at Rockland, located on the Neches River, of 67 in. The unusual east prevailing wind caused a westward curving of the 55 and 50-in. isohyets. The concentration of rainfall in Montgomery County was likely due to persistent stalling of frontal activity north of Houston. The year, 1929, although near normal at the downtown recording station, was preceded by two below normal years, and followed by five years below normal.[41, 42, 43]

The 1935 map, with a maximum of 66 in. at Conroe, shown in Figure 25,

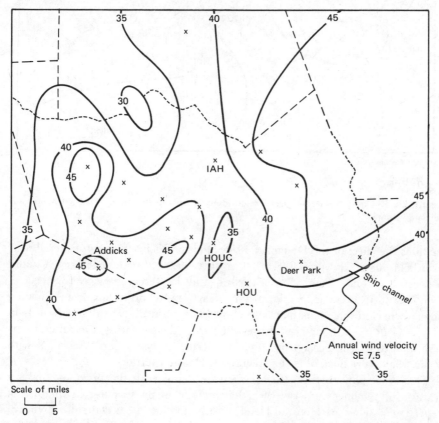

Scale of miles

0   5

Figure 19. Annual precipitation in 1971 for Houston metropolitan area, Harris County, and adjoining counties.

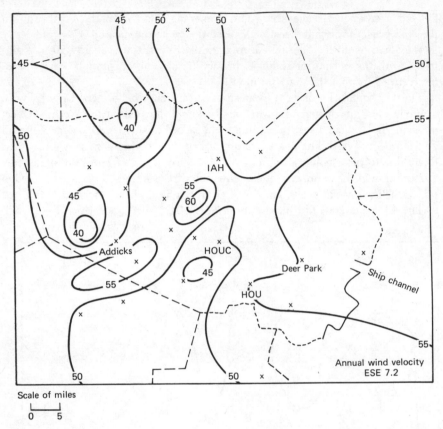

Figure 20. Annual precipitation in 1972 for Houston metropolitan area, Harris County, and adjoining counties.

contains the effect of the December 1935 Houston Flood that was centered near Westfield along Spring Creek and in the western half of Harris County between Katy and Satsuma[43]. Barker and Addicks Reservoirs were constructed, following this disaster, to keep the heavy rains upstream from flooding downtown Houston. This map also shows lack of available rainfall records in Harris County. Although wind prevailed from the southeast, the rainfall pattern indicates an increase from south to north, similar to the 1929 map, with about 5 in. above normal at the downtown Houston station.

   The 1946 map (Figure 26) contrasts in detail with the previous analyses because of the increased network density due to an increase in number of reporting stations. At Houston Heights 78 in. fell just a few miles downwind of the downtown station where 69 in. fell. And 71 in. fell at Tomball, northwest of Houston, with lighter rainfall to the southwest, at Katy, and a maximum eastward in the Baytown (Goose Creek) area. Note the range from 61 in. at Katy to 71 at Tomball and 86 in. at the extreme east end of the county. This map illustrates the cellular pattern, which is evident in the dense micronetwork of stations. The wind was from the southeast, and the general pattern of wet in the East, and dry in the west prevailed. However, the cellular pattern is apparent west of the downtown area and northwest near Tomball. This year was in a wet period, with 69 in. downtown, preceded by 56 in. in 1945, and followed by only 41 in. in 1947.

The 1958 map (Figure 27) illustrates the pattern that prevails during a year of slightly below normal rainfall. The area less than 35 in. covers a large portion of western and northeastern Harris County. A reason for the 10-in. increase in rainfall between Satsuma and Cypress is not immediately apparent and will require additional research. A south-southeast wind prevailed, and the

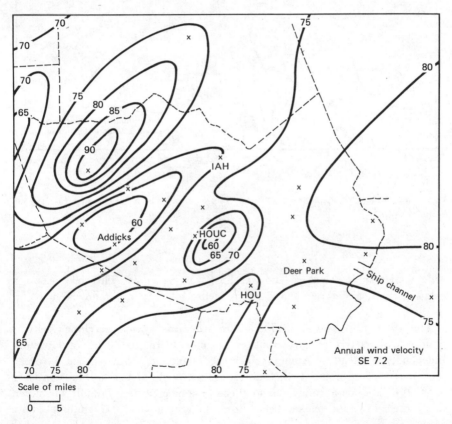

Figure 21. Annual precipitation in 1973 for Houston metropolitan area, Harris County, and adjoining counties.

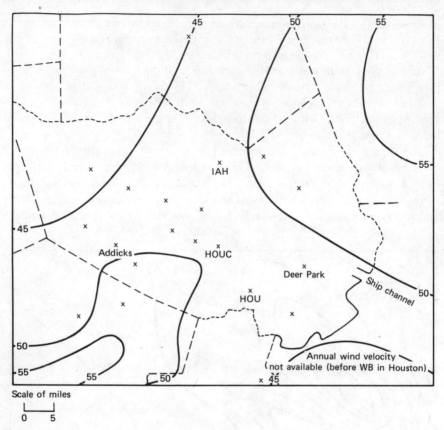

Figure 22. Annual precipitation in 1908 for Houston metropolitan area, Harris County, and adjoining counties.

general pattern of dry West and wet East was preserved. This year was in a below normal rainfall period and was preceded by 57 in. and followed by 64 in. in 1959.

The 1966 map, shown in Figure 28, contains a well developed cellular pattern with cells that vary from 48 in. downtown, to 29 in. at Houston-Independence Heights. Amounts increased westward to 54 in. at Cypress, southwestward to 57 in. at Alief, and southeastward to 64 in. at Houston-Deer Park. With the wind from south southeast, the near normal rainfall downtown was preceded by four years of below normal and followed by an amount of only 30 in. in 1967.

Some Results of the Houston Study

The transition from a generally linear rainfall pattern to a cellular pattern shown to occur in the various spatial distributions of annual rainfall seems to be permanent. The predominance of a dry cell over the urban center with downwind maxima during both wet and dry years seems to be a permanent feature of the Houston metropolitan area. The location of the maxima downwind of the tall structures rather than confined to downwind of the Ship Channel industrial complex portends that the principle enhancement of rainfall is by increased convection by obstructions to airflow. The examples of the rainfall changes at two stations, located along the average annual wind direction,

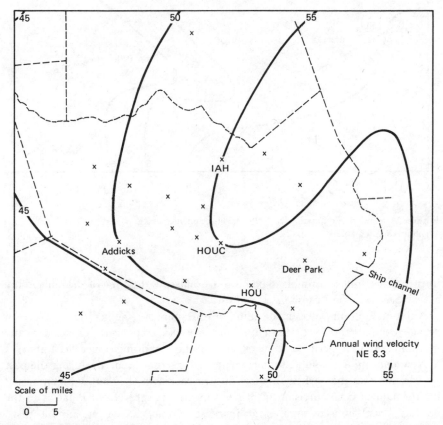

Figure 23. Annual precipitation in 1913 for Houston metropolitan area, Harris County, and adjoining counties.

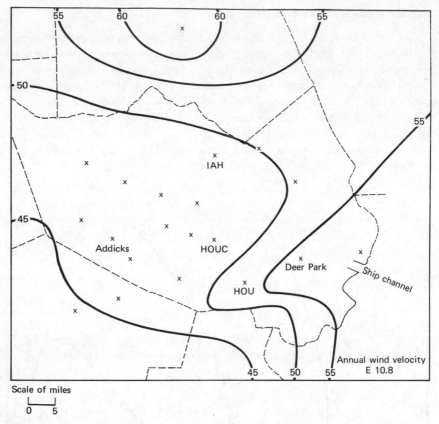

Figure 24. Annual precipitation in 1929 for Houston metropolitan area, Harris County, and adjoining counties.

illustrate the shift in rainfall maximum and narrow the time of the shift to the 5-year period of most rapid urbanization.

An initial inquiry was made into the question of whether the increase in decade rainfall applies to the entire upper coast area or is restricted to the metropolitan area, perhaps at the expense of the remaining upper coast area.

New data, now available to document changes that occurred during the past decade, indicate that, while the new 30-year normals of monthly precipitation for the Upper Coast division of Texas (an area roughly from Palacios to the Sabine River) indicate an average increase of only 1% on an annual basis, monthly values vary from as low as 87% of normal in July to as high as 121% of normal in June. There is a marked increase of 30% (10% below normal to

20% above normal) from March to June. This change is particularly significant since this period coincides with the spring flood season. A similar increase of 26% (13% below normal to 13% above normal) from July to September coincides with the hurricane season. This monthly distribution is vital to allow for increased flood discharge at both spring and autumn seasons. These changes are illustrated in Figure 29, while annual trend changes from 1941 to 1972 for the Upper Coast are shown in Figure 30.[36, 37]

A comparison of the 30-year normal ending in 1970 versus the normal ending in 1960 for Intercontinental Airport is illustrated in Figure 31. By comparing this figure with Figure 29, which is a similar distribution for the Upper Coast, we see that the distribution from July through September is flatter for

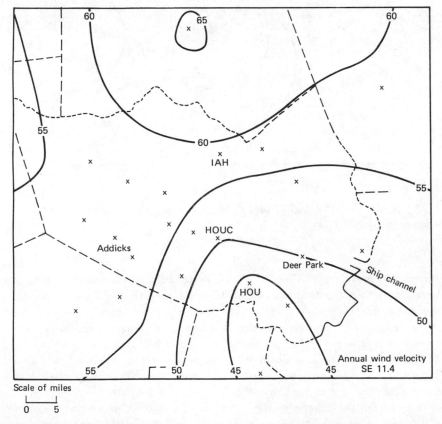

Figure 25. Annual precipitation in 1935 for Houston metropolitan area, Harris County, and adjoining counties.

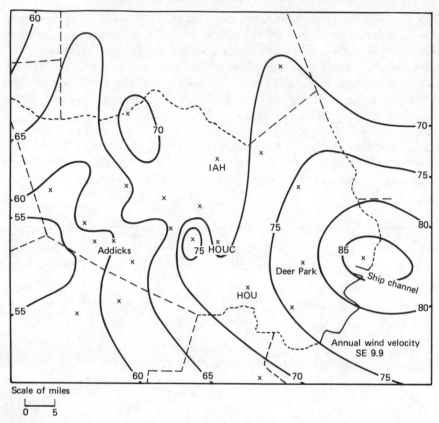

Figure 26. Annual precipitation in 1946 for Houston metropolitan area, Harris County, and adjoining counties.

Intercontinental Airport. This is to be expected because the tropical storm and hurricane season causes the peaked distribution, and the coastal area responds more to tropical storm occurrence than does the airport, located 70 mi. inland from the coast. The new 30-year normal at the airport ending in 1970 increased by 5%, while the like normal for the Upper Coast increased by only 1%. This higher percentage implies that precipitation north of downtown Houston, and downwind of the urban industrial complex, has increased more in the past decade than the Upper Coast division as a whole.

A tool from the Houston study[7] that may be useful for determining urban trends elsewhere is the 10-year moving average. Fortunately, waiting until the end of a decade is unnecessary to reveal expected changes in a trend when a 10-

year moving average is monitored. Because of its proximity to the annual normal in rainfall amount, the transition month of October was chosen to illustrate the annual normal. Averages were computed for 10 consecutive Octobers and plotted using the record period from 1940 to 1970. The general precipitation trend between 1949 and 1970 can be determined for the month of October from the results of the plot of the 10-year moving average. This is illustrated in Figure 32. Because of the nature of this index, predictions are usually made toward conserving the normal. With the trend falling between 1958 and 1970, a sharp rise, such as occurred in Houston from 1971 to 1973, was expected.

It remains to be seen whether variations from the normal will continue in this trend or will grow as did the decade averages for annual rainfall. The

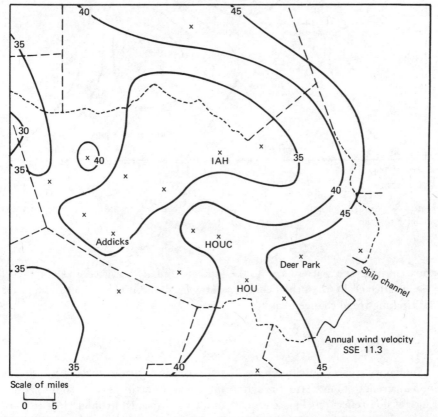

Figure 27. Annual precipitation in 1958 for Houston metropolitan area, Harris County, and adjoining counties.

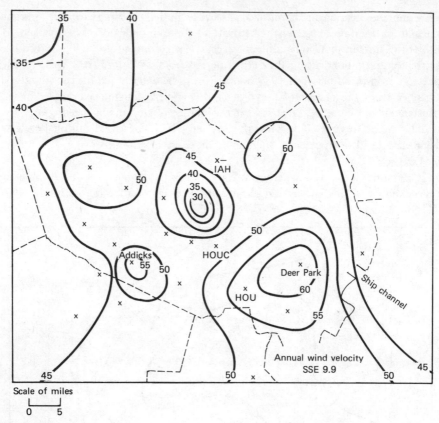

Figure 28. Annual precipitation in 1966 for Houston metropolitan area, Harris County, and adjoining counties.

apparent differences are related to the type of rainfall characteristics of the Fall season, affected by the urban complex. However, the usefulness of the tool for anticipating trend change is quite apparent.

Conclusions

Humanity's effect on climate has been increasing. Just as history has shown human strength to increase through forming groups acting as one, so the significance of their effect on climate grows as they have formed the high density complex called the urban environment. The increase is increasingly apparent as more data are concentrated near urban areas. While empirical evidence of climatic change is mounting, study efforts are directed at particular mechanisms of climatic changes caused by urbanization. Climatic changes are being revealed worldwide at a faster rate than are the studies to analyze the evidence. The few studies that have been initiated indicate the importance of considering contemporary effects on climate through urbanization and the implications of these effects.

The importance of responding to evidence of climatic changes early, should be recognized and acted upon with directed activity focused on the preservation of the natural environment. This is done easily by moving in the direction of

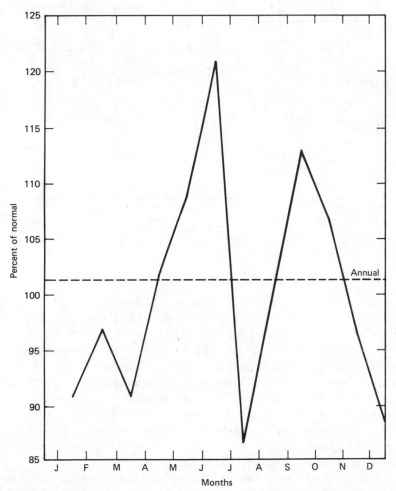

Figure 29. Distribution of monthly rainfall for Upper Coast of Texas in percent of 30-year normal 1931 to 1960. Dashed line is new 30-year normal 1941 to 1970.

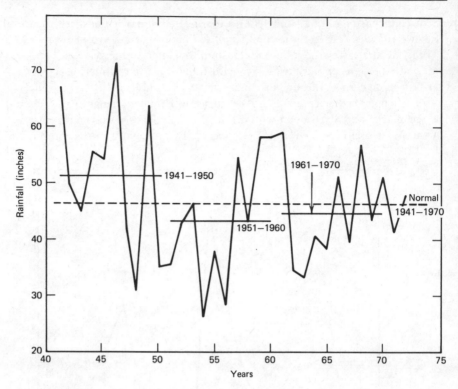

Figure 30. Annual rainfall Texas Upper Coast for record period 1941 to 1972. Decade average rainfall indicated by horizontal lines.

least resistance, the direction of natural environmental forces. One apparent natural direction is the prevailing wind direction. Examples are using the direction of the wind to locate future areas for "water catchbasins" downwind of moisture sources, and locating reservoirs fed from subsurface upwind of the urban area, in the direction of dry air sources, to help modify those air currents that would ordinarily be hot and dry. City planning should include locating industrial heat sources in a portion of the metropolitan area where they can enhance desirable climatic effects for both the rural and urban environment. These concepts were proposed in detail by Dr. Geoffrey Stanford, who, in addition, has conceived of using land tilling techniques to enhance agriculture in semi-arid areas that would otherwise be unproductive.[40]

The results of the initial study of Houston's change in rainfall pattern should be a warning for other megalopoles worldwide. If the spatial rainfall gradient can be reversed in Houston, what can happen to other fast growing urban centers? Are there evident trends that have been overlooked in lieu of the more

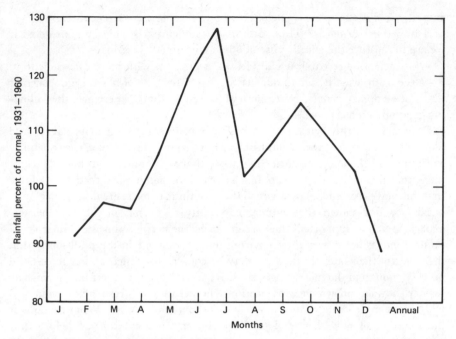

Figure 31. Distribution of monthly rainfall for Intercontinental Airport, Houston, Texas. Normal comparison as a percent of 1931 to 1960 normal. Percent increase (5%) of new normal for 1941 to 1970 indicated by horizontal line in annual column.

visible, oppressive, audible, and olfactory results of air, water, noise, and visual pollution? How are the known implications affecting the economic potential of urban centers? How is the changing roughness of the boundary layer in the urban expanse affecting climate, adversely or beneficially? Can some of the adverse effects be made beneficial for future growth? In particular, can we

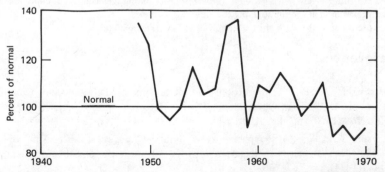

Figure 32. Houston precipitation (10-year moving average), month of October 1949 to 1970.

expect to build effectively to help extract sufficient water from the air to support future urban needs? These are all questions that should be considered if we are to improve the effective use of our environmental resources.

While scientific research usually leaves a problem with more questions than it started with, usually due to our arising from the state of blissful ignorance, it may not be appropriate to conclude with only questions. Therefore, the following generalization is made:

Contrary to the cries of various environmental alarmists (Kay and Skolnikoff,[47]) people, throughout history, have shown flexibility in the capability to circumvent the many obstacles in nature that have confronted them. It seems likely that they will have the capability as well to change the direction of a force that they have created and over which they continue to exert great influence.

History has shown that human capability to reason in the face of new experiences has supported their reign over their environment. The role of humankind as ruler over their environment is inherent in the philosophy that they control their fate. While there may be concern over their ability to respond quickly enough to the increasingly sensitive environment to avert inconvenience, history shows us no evidence to question whether they will respond to avoid ultimate disaster. It is not a question of will humankind survive. What remains is the question of how degraded they let the environment become before they respond.

Eventually urbanization effects on climatic change, along with other results of human manipulations of the environment, must be directed *with* rather than *against* nature, toward beneficial modification of climate.

Acknowledgments

It is with great pleasure and pride that I acknowledge Mr. Clayton B. Crooker for initially alerting me to the Houston climatic change and for his untiring efforts during the ensuing data collection and analysis. I am indeed proud to be his colleague and very grateful for his support.

Special recognition is given Ms. Danila Perossa for her encouragement and support during the manuscript preparation and for applying the finishing touches.

The perspective art is the work of Mr. Steve Knight, an exceptional talent.

References

1. Altshuller, A. P., "Composition and Reactions of Pollutants in Community Atmospheres," Urban Climatology, *World Meteorological Organization Technical Note 108* (1970), pp. 179–193.

2. Bryson, Reid A., "A Perspective on Climatic Change," *Science,* Vol. 184, No. 4138, May. American Association for the Advancement of Science Washington, D.C. (1974), pp. 753–760.

3. Changnon, S. A., Jr., "The La Porte Weather Anomaly—Fad or Fiction?" *Bulletin of the American Meteorological Society,* Vol. 49 (1968), pp. 4–11.

4. Changnon, Stanley A., Jr., "Inadvertent Weather Modification," Eleventh American Water Resources Conference, November 10–13, 1975, Baton Rouge, La.

5. Charney, Jule, and Stone, Peter H., "Drought in the Sahara: A Biogeophysical Feedback Mechanism, *Science,* Vol. 187 (1975), pp. 434–5.

6. Crooker, Clayton B. and Goldman, Joseph L., "Preliminary Analysis, Flood of March 20, 1972," Institute for Storm Research, Houston, Texas, 1972.

7. Crooker, Clayton B. and Goldman, Joseph L., "The Changing Climate, An Analysis of Rainfall and Temperature for Houston-Galveston and 13 County Areas 1901–1973," Institute for Storm Research, Houston, Texas, 1974.

8. Crooker, Clayton B. and Goldman, Joseph L., "Climate Modification in an Urban Metropolitan Area," 55th Annual Meeting of the AGU, April 8–12, Washington, D.C., 1974.

9. Crooker, Clayton B. and Goldman, Joseph L., "The Changing Mesoclimate, Further Evidence of Anthropogenic Effects in Houston," AGU Fall Annual Meeting, December 12–17, San Francisco, Cal., 1974.

10. Fedorov, Y. K., "Modification of Meteorological Process," *Weather and Climate Modification,* New York: John Wiley, 1974, pp. 387–401.

11. Fischer, William H., and Sturdy, G. C., "The Contribution of a City to Atmospheric Turbidity and the Turbidity Background," *Atmospheric Environment* Vol. 5 No. 7 (1974), pp. 561–563.

12. Frisken, William R., "Resources for the Future, Inc.," *The Atmospheric Environment,* Baltimore, Md.: John Hopkins University Press, 1973.

13. Haber, George, "The Crumbling Shield," *The Sciences,* New York: New York Academy of Science, 1974, pp. 21–24.

14. Hammond, Allen L., and Maugh, T. H., II, "Stratospheric Pollution: Multiple Threats to Earth's Ozone," *Science,* Vol. 186, No. 4161 (1974) pp. 335–338.

15. Harnack, R. P. and Landsberg, H. E., "Selected Cases of Convective Precipitation Caused by the Metropolitan Area of Washington, D.C.," *Journal of Applied Meteorology,* Vol. 14 (1975) pp. 1050–1060.

16. Hiatt, William E., "The Analysis of Precipitation Data," The Physical and Economic Foundation of Natural Resources Series, U.S. Weather Bureau, Washington, D.C., 1951.

17. Huff, F. A. and Changnon, S. A., "Precipitation Modification by Major Urban Areas," *Bulletin of the American Meteorological Society,* Vol. 54, No. 12 (1973), pp. 1220–1232.

18. Johnson, S. L., "Urban Hydrology, Houston Metropolitan Area, Texas," U.S. Geological Survey Open-File Report. (1966 to 1969)

19. Johnson, S. L., and Sayre, D. M., "Effects of Urbanization on Floods in the Houston, Texas Metropolitan Area," U.S. Geological Survey, Water Resources Investigations, 1973, pp. 3–73.

20. Jones, Roscoe H., "Major Office Areas 1960–1990," Houston City Planning Department, City of Houston, Texas, 1971.

21. Kellogg, W. W. and Schindler, S. H., "Climate Stabilization: For Better or For Worse?" *Science,* Vol. 186, No. 4170 (1974) pp. 1163–1172.

22. Landsberg, Helmut, "The Meteorologically Utopian City," *The Bulletin of the American Meteorological Society,* Vol. 54, No. 2 (1973), pp. 86–89.

23. Landsberg, Helmut, "Inadvertant Atmospheric Modification through Urbanization," In W. N. Hess, Ed., *Weather and Climate Modification,* New York: John Wiley, 1974.

24. Leipper, Dale F. and Volgenau, D., "Hurricane Heat Potential of the Gulf of Mexico," *Journal of Physical Oceanography,* Vol. 2, No. 3 (1972), pp. 218–224.

25. Lowry, William P., "The Climate of Cities", *Scientific American,* Vol. 217, No. 2 (1967) pp. 15–23.

26. Machta, Lester and Telegadas, Kosta, "Inadvertent Large-Scale Weather Modification," In *Weather and Climate Modification,* W. N. Hess, Ed., New York: John Wiley, 1974, pp. 687–725.

27. Matthews, William H., Kellogg, W. W., and Robinson, G. D., Eds. *Man's Impact on Climate,* Cambridge, Mass.: MIT Press, 1971, p. 594.

28. MIT, *Inadvertent Climate Modification: Report of the Study of Man's Impact on Climate,* Cambridge, Mass: MIT Press, 1971, p. 309.

29. Namias, Jerome, "Macroscale Variations in Sea-Surface Temperatures in the North Pacific," *Journal of Geophysical Research,* Vol. 75, No. 3 (1970).

30. Namias, Jerome, "Collaboration of Ocean and Atmosphere in Weather and Climate," Proceedings of the 9th Annual Conference of the Marine Technology Society, Washington, D.C., 1973.

31. Namias, Jerome and Born, R. M., "Further Studies of Temporal Coherence in North Pacific Sea Surface Temperatures," *Journal of Geophysical Research,* Vol. 79, No. 6, (1974).

32. Newell, Reginald E., "The Earth's Climatic History," *Technology Review,* Cambridge, Mass.: MIT Press, 1974, pp. 31–45.

33. Newman, James E. and Pickett, R. C., "World Climates and Food Supply Variations," *Science,* Vol. 186, No. 4167, (1974), pp. 877–881.

34. NOAA, Environmental Data Service, "Climatological Data—Texas," Asheville, N.C.

35. NOAA, *Climatology of the United States,* No. 81 (by state), "Monthly Normals of Temperatures, Precipitation, and Heating and Cooling Degree Days 1941–70," Environmental Data Service, National Climatic Center, Asheville, N.C., 1973.

36. NOAA, *Local Climatological Data,* Environmental Data Service for Galveston and Houston, Texas.

37. NOAA, *Monthly Averages of Temperature and Precipitation for State Climatic Divisions 1941–1970, Texas,* Environmental Data Service, National Climatic Center, Asheville, N.C., 1973.

38. NOAA, *Climates of the States, Texas, Climatography of Texas,* "Normal Annual Total Precipitation (Inches)," Environmental Data Service, Asheville, N.C., 1968.

39. Roberts, Walter Orr, "Possible Relationships Between Solar Activity and Meteorological Phenomena," *Transactions, American Geophysical Union,* Vol. 55, No. 5 (1974), pp. 524–527.

40. Stanford, Geoffrey, *The Handbook of the Odessa Project for Resources Recovery,* Environic Foundation, International, University of Notre Dame, Indiana, 1974.

41. U.S. Weather Bureau, *Climatography of the United States,* Climatic Guide, Houston-Galveston, Texas Area, No. 40–41.

42. U.S. Weather Bureau, *Climatic Summary of the United States,* Bulletin W, Section 33, 1930.

43. U.S. Weather Bureau, *Climatic Summary of the United States, Texas Supplement for 1951–1952.*

44. U.S. Weather Bureau, *Climatic Summary of the United States, Texas Supplement for 1951–1960.*

45. Water Resources Data for Texas, Part I, *Surface Water Records,* U.S. Geological Survey, Austin, Texas, 1970–1971.

46. Williams, G. R., *Engineering Hydraulics,* H. Rouse, Ed., New York: John Wiley, 1949, pp. 229–320.

47. Kay, David A., and Skonikoff, E. B., Eds., *World Eco-Crisis,* University of Wisconsin, Madison, Wisconsin.

4

Social Impacts

of Climate Modification

J. EUGENE HAAS
Department of Sociology
Institute of Behavioral Science
University of Colorado
Boulder, Colorado

Over the centuries man has developed more and more technologies to cope with day to day and seasonal variation in the atmosphere around him. His buildings, heating and cooling devices, dams and levees, drought-resistant crops, drainage and irrigation systems, "weatherproof" modes of transportation, and weather forecasting techniques all make him less susceptible to the vagaries of nature. They allow him to live and work, even thrive, in what can reasonably be called inhospitable climatic conditions.

But these adjustments have yet to make human activity completely immune to the temperature, precipitation, and wind parameters of the atmosphere.[16] Snow and ice are still very significant impediments to air and highway transportation. And wind, ice, and snow can still turn hundreds of thousands of warm homes and offices into something akin to refrigerated storage areas. Man's increasing success in controlling the temperature in his buildings also represents his increasing vulnerability to the failure of his electric power system when he has no readily available substitute technology.

Interaction Between Human Activity and the Climate

On the whole, however, when modern mans' technology is functioning as it was designed to do, weather and climate have little influence on his activities.

103

Or do they? Even in modern society much of man's activity still seems to be weather and climate related. Most people still make summer their vacation period. Public meetings often have reduced attendance on rainy days. Schools and businesses by the hundreds cease operations during and following a snow storm. Hospitals and automobile insurance adjustors have an upswing in activity when there is rain or snow following an extended dry spell. Children play outside much more on warm, dry days than on cool, wet ones. On warm days beaches are jammed; on cool days, deserted.

Clearly, temperature and precipitation still do make a difference in the intensity and characteristics of many human activities.

Assumptions and Limitations

The analysis and the generalizations in this chapter are intended to apply to the United States. The generalizations may also apply to Canada, but it is unlikely that they are generally valid for all societies. At least there is no good evidence to support that view.

This section is based on a set of assumptions about supersonic transport induced climatic change.[6] They include the following:

1. The maximum credible temperature change will be a mean annual decrease of 2.5°C.
2. The maximum credible precipitation change will be an annual increase or decrease of 20%.
3. When the maximum precipitation increase occurs there will be fewer days with no precipitation throughout the year, and there will be an increase in the number of days in a year with precipitation of one inch or more.
4. Conversely, when the maximum precipitation decrease takes place, there will be more days with no precipitation and fewer days when one or more inches of precipitation occur.
5. The combination of temperature decline, a decrease of as much as 2.5°C, and up to 20% increase in precipitation, will produce more ice storms and a higher proportion of snow to rain from total yearly precipitation.
6. With only minor exceptions, there will not be any major changes in technology before 1990, significantly decreasing the impact of temperature and precipitation on man's principal activities.

Extension of the Winter Season

Longer and colder winters also mean cooler and shorter summers. While the short-term fluctuations in temperature will still be present, the mean seasonal temperature will decline. One way to get a better grasp of what this means for human activity is to consider the following temperature changes by city within the contiguous 48 states. After climatic change, cities listed on the left will have winters and summers much like the cities listed on the right *now* have.

Kansas City	Omaha
Omaha	Sioux Falls
Sioux Falls	Minneapolis
Minneapolis	Duluth
Nashville	Louisville
Louisville	Cincinnati
Cincinnati	Fort Wayne
Shreveport	Little Rock
Little Rock	Springfield, Mo.
Springfield, Mo.	St. Louis
Pueblo, Colo.	Denver
Denver	Cheyenne, Wyo.

These pairings of before and after cities are, of course, only approximate. If the temperature decline is accompanied by as much as a 20% *decrease* in precipitation then the pairings may overestimate the significance of the climatic change. Cooler days without precipitation produce less change in activity than cooler days with precipitation. If the precipitation does increase up to 20%, the pairings may underestimate the significance of the change. This would apply to both summer and winter.

Longer, colder, and wetter winters mean more sports activities dependent on snow and ice. They also mean higher snow removal costs,[1, 5] more time on the road for basic activities such as getting to work and shopping, more weeks during which automobiles have studded snow tires that bring more deterioration to street and highway surfaces, more heaving of road surfaces requiring additional repair, more salt on the streets and highways to rust out the bottoms of cars faster, as well as more road service for stalled and disabled vehicles.

Shorter, cooler, and wetter summers mean fewer hot, humid, and sleepless nights but also increased effort to protect oneself from more mosquitoes. Such summers mean less time spent watering lawns and gardens but also higher food costs due to increased transportation of fruits and vegetables that require longer growing seasons. They mean fewer crops plowed under or used for forage due to lack of precipitation but also more crops lost at harvest time due to muddy soil unable to support the heavy harvesting machinery. Such summers also mean more time and money washing the car; more picnics and rodeos interrupted or postponed by rain; more "miserable" camping trips; fewer forest, brush, and grass fires; and, of course, fewer comfortable hours around the swimming pool and at the beach.

It should be clear already that there will be changes in activity levels growing out of climatic change. There will be some modest and some significant alterations in life style. For most persons some of these changes will be seen as desirable and others as definitely undesirable. Few persons enjoy being cold and damp or wearing the necessary clothes to assure that the rain and snow won't make them so. Most persons won't complain about experiencing fewer excessively hot nights. Almost everyone would prefer to avoid the higher costs of snow and ice removal. Some persons enjoy skiing, snowmobiling, and ice skating, but most Americans do none of these.

Extending the winter season will bring a range of changes. They are discussed in detail later in the chapter.

There appears to be a trend toward year-round operation of elementary and secondary schools. Under this arrangement about an equal number of students are out of school during each month of the year. The potential is increasing, then, for family vacations to be taken in almost any month of the year. Many farmers already take vacation in the winter. How will these "out-of-town" vacations be affected by the anticipated climatic change? If the entire country is 2 to 2.5°C colder in the winter, will there still be a shift southward in the principal winter vacation spots for Americans? The answer is probably in the affirmative.

It seems to be true that within limited ranges the human response to temperature is relative. At 75°F Montanans think it is very warm, while most Floridians describe the temperature as rather cool. Nevertheless, when winter comes and tens of thousands of Americans go southward to escape the cold, most of them go far enough south to be comfortable in a swimming suit. And for most of them that means going perhaps an additional 200 miles south. Precisely where they go depends on many factors in addition to the temperature, but a person who plans to go 1000 to 1500 mi to the south is not likely to hesitate long about going an additional hundred miles or so to get to where it is "really warm." Southern Florida, the Caribbean area, the south Texas coastal area, and Mexico are likely to become even more the winter vacation spots of the "average" American family than is currently the case.

If, indeed, that revised "Southward Ho" vacation pattern gets established, there will be a number of consequences flowing therefrom: more time and money spent traveling to and from the winter vacation spot; a larger proportion of the family budget spent on vacation; and an increasing proportion of the nation's fuel being used on winter vacations.

Colder and Drier or Colder and Wetter

Atmospheric Hazards

First consider how changes in climate may affect the occurrence and intensity of those major natural hazards closely related to the characteristics of the climate.

The best available evidence suggests that direct economic loss due to flooding in the United States runs on the order of $1.5 billion per year.[15] For losses due to snow the figure is $15 million.[5] A 20% decrease in precipitation would certainly bring some reduction in such economic losses and social disruption. A conservative figure would be a 10% decrease. In strict economic terms that would represent losses avoided of $150 million and $1.5 million for flooding and snow storms respectively.

If that same conservative 10% change figure is applied to the condition of precipitation increase, then stratospheric transport would produce additional annual economic losses on the order of $150 million from flooding and $1.5 million from additional snow. A significant increase in social disruption (e.g., school days lost, trips canceled) would also take place. See Figures 1 through 4 for relevant data.

When hurricanes strike the Gulf and Atlantic Coasts of the United States they produce massive social disruption. It is not unusual for several hundred thousand people to have to evacuate coastal areas in the face of an approaching hurricane. Quite apart from the direct storm produced damage, this social disruption must also represent a very significant economic loss. And the additional fuel consumed in the evacuation and return is not inconsequential either.

This author knows of no serious effort to link the assumed climatic changes with the frequency, intensity, and storm tracks of hurricanes that could impact the United States. If we assume that the characteristics of the hurricanes will remain unchanged, what are the implications? If the climatic change includes a significant decrease in precipitation along the Gulf and Atlantic Coasts, the massive rainfall from the hurricane would produce less local flooding than is now the case.[14] The soil should be able to absorb more of the moisture, leaving less for fast runoff and flooding. If, on the other hand, there is a 20% increase in precipitation, the rains associated with the hurricane are likely to fall on saturated or nearly saturated soil, and damage from flooding would be sharply increased.

Given current practice in recording economic losses due to hurricanes, it is not possible to separate out in any very precise manner losses due to inland flooding from losses due to storm surge and/or wind.[2] However, we do have available estimates that mean annual losses from hurricanes in general are cur-

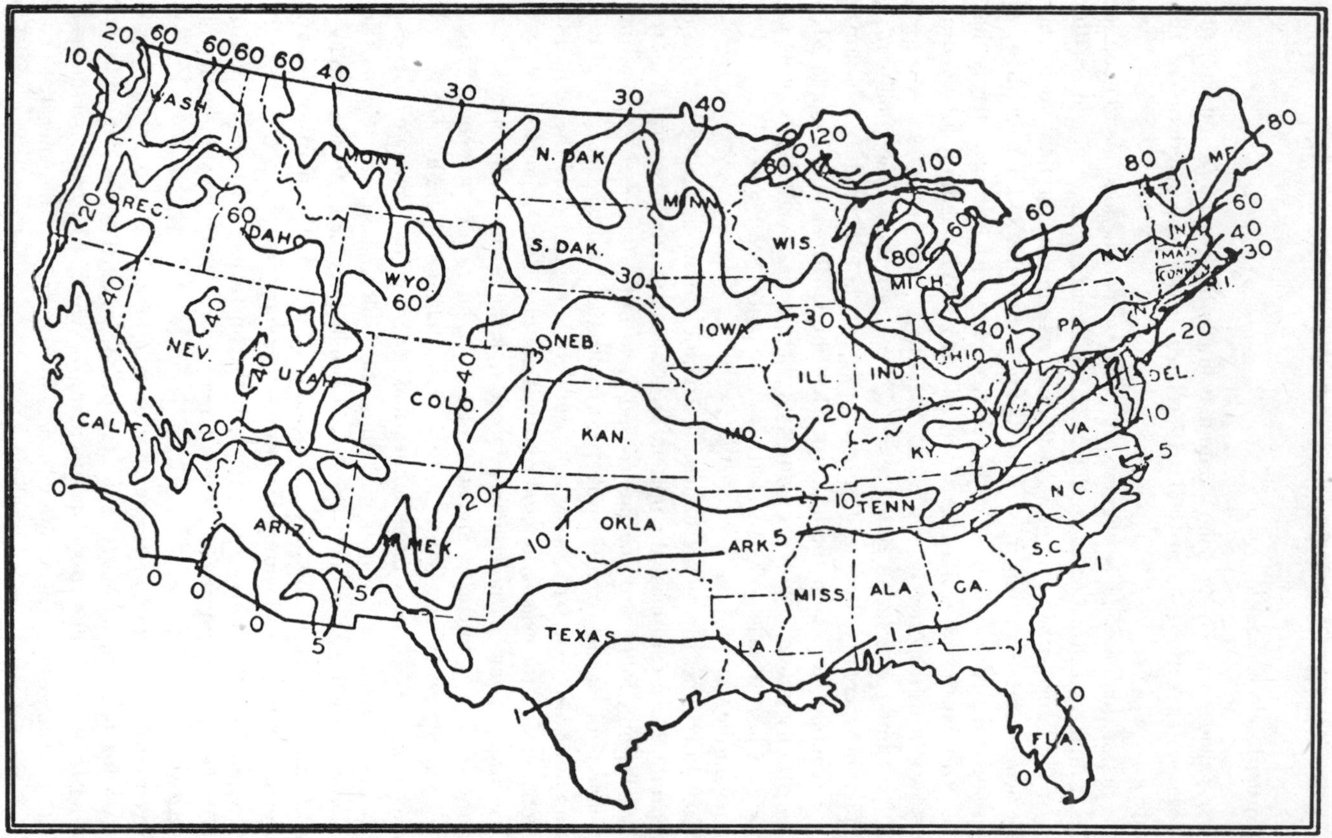

Figure 1. Average total snowfall in inches per year.[1]

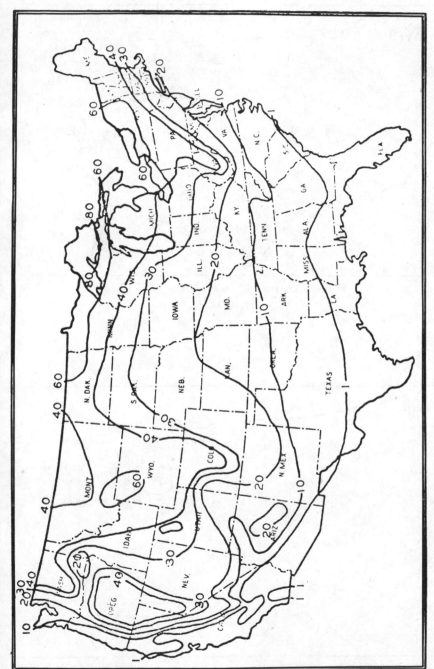

Figure 2. Average annual number of days with a snowfall of 0.01 in. or more.[1]

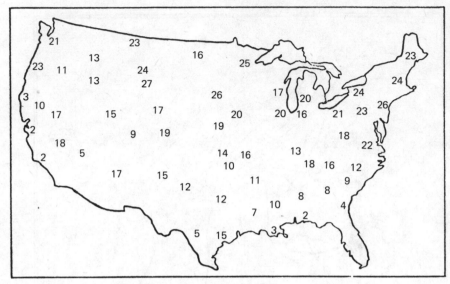

Figure 3. Extremes of 24-hour snowfall in inches.[14]

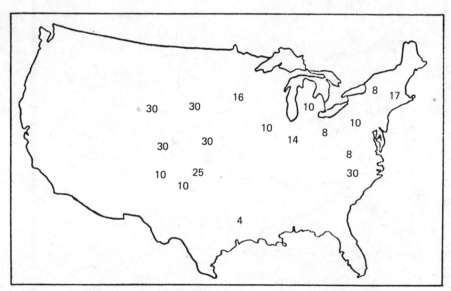

Figure 4. Heights of snow drifts in feet after some outstanding snowstorms of this century.[13]

110

rently running about $450 million.[2] We would guess that figure could be altered by about 2%, or $9 million, annually in either direction due to decreased or increased inland flooding resulting from climatic change. Any changes in the level and character of social disruption would be minor.

Hail storms and tornadoes are principally a product of continental rather than maritime weather systems.[17] The dynamic processes that produce damaging hail and tornadoes are not well understood. Forecasting with much precision where the damage will occur is also at a primitive level. However, there are relatively adequate data on crop damage due to hail and property damage (mostly buildings), due to tornadoes. These average annual losses respectively are $500 million[3] and $100 million.[4]

Storms that produce hail and tornadoes occur in every one of the 48 contiguous states, but there are major differences in frequency (see Figure 5). In general, the Pacific and Atlantic coastal states have lower frequencies, and the extreme northern states have fewer such storms than do those further south. These storms develop and feed on warm air and moisture. Thus, if the climatic change is one of lower temperatures and less precipitation, any effect on severe local storms is likely to represent a decrease in damaging hail and tornadoes. Lower temperatures and more precipitation might be expected to produce no

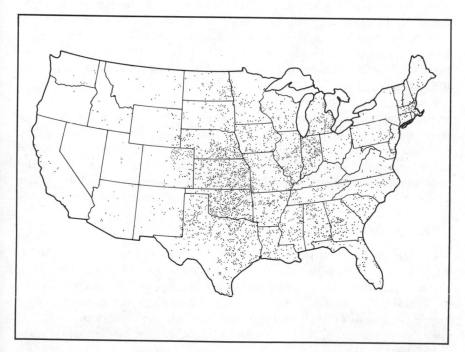

Figure 5. Tornadoes, 1953 to 1958.[17]

significant change or perhaps a slight increase in hail and tornado frequency. However, the frequency of the hail or tornado event taken alone is not what is critical. If the climatic change produces significant alterations in cropping patterns throughout the United States, total hail-caused damage could go sharply upward or downward. A crop such as tobacco has a much higher value per acre than does wheat, for example. The total damage level is very largely dependent on cropping patterns. In the absence of estimates of changes in such patterns it is unrealistic to try to estimate what loss trends for hail damage are likely to be.

Significant tornado damage is a function of the location, strength, and use of man-made structures. Recent and current trends in population mobility, and therefore man-made structures, is toward the south and west. This trend probably increases slightly the vulnerability of structures to tornado damage. If, as a result of climatic change, the areas of high tornado frequency move southward, then population mobility trends (if they continue), will coincide with changing tornado frequency trends to produce significantly larger annual tornado losses. A 5% increase in annual losses to tornadoes would represent about $5 million.

Crop losses due to frost are estimated to be between $120 million[9] and $1 billion[11] yearly. If it is assumed that cropping patterns will gradually shift to take the decrease in frost-free days into account, then the direct frost losses should not be appreciably altered. However, unless there is a very rapid and widely publicized shift in the number of frost-free days for any given area, it must be concluded that the change to a cooler climate will bring with it sharply increased losses to frost for perhaps 10 to 15 years. Basic ideas about farming do not change quickly.[11]

Losses to snow avalanches will increase or decrease depending on changes in precipitation levels. However, since the temperature is assumed to decline, a decrease in precipitation is likely to bring only a slight decrease in avalanches, while an increase in precipitation accompanied by lower temperatures would almost certainly bring a sharp increase in avalanches. Given the rapidly increasing popularity of skiing and especially cross-country skiing, deaths due to avalanches and/or blizzards are likely to show a dramatic upturn.

Changes in Family and Community Activity

Perhaps the best way to grasp the implications of climatic change is to list the range of people's activities and try to specify how those activities would be altered, if at all, by the anticipated changes in temperature and precipitation. By thinking of a typical United States community and the families residing therein, it is possible to be relatively specific about the changes that will take place. Table 1 displays, in abstracted form, what seems to be the likely changes or nonchanges.

Table 1. Outline of Anticipated Changes in Family and Community Activity Due to Two Types of Climatic Change

Activity	Average Weather Conditions	
	Colder and Wetter	Colder and Drier
LOCAL		
Within the nuclear family	Increased adult-child interaction Increased sibling interaction	Less significant change in same variables
Neighboring	Reduction in frequency	Minor change only
Journey to work:	Increased tardiness and absenteeism More time spent on road	No significant change
Shopping	Slight increase in time per unit purchased	No significant change
Education	More school closings Increase in unsupervised children at home Reduction in outdoor exercise for children	Minor change only
Leisure and recreation	Sharp reduction in outdoor activity (winter) Significant reduction in outdoor activity (summer) Increase in indoor, passive leisure	Slight reduction in outdoor activity (winter) Slight increase in outdoor activity (spring, summer, and fall)
Health care	For both home and hospital, increase in time spent caring for sick and injured	No significant change
Political activity	Reduced citizen participation Additional reliance on TV during political campaign	Minor change only
Other business activity	Greater day-to-day variance in volume of retail sales More frequent delays in delivery of products and services	Minor change only
Religious activity	Reduction in participation in organized activity	No significant change
DISTANT		
Familial	More interrupted and cancelled trips	No significant change
Leisure and recreation	Increase in interrupted and cancelled trips (winter) Decline in trips to certain areas (all seasons)	Minor change only
Commuting to work	Decline in number of commuters using autos	No significant change

In this portion of the text we note only those activities for which there is anticipated change. The abbreviations to be used are C and W for "colder and wetter" and C and D for "colder and drier."

Within the Nuclear Family.   When it is C and W, both children and adults tend to stay indoors more, especially in the winter months. Thus, there is an increase in face-to-face interaction. Within certain limits this will produce no negative effects such as heightened irritation and quarrels. Such limits vary by family and within the same family will vary over time.

With children spending less time outdoors there will be increased interaction between parents and their children. Thus, in the socialization process for children there will be relatively more parental and sibling influence (or at least more opportunity for such influence), and less peer influence than would otherwise be the case. There will be more sibling play and more sibling fighting. Members of the same household are less likely to be strangers to one another.

Under C and D conditions the same changes will take place but only to a limited degree. It is not too difficult to dress for somewhat colder weather providing you don't have to contend with precipitation much of the time. Thus, children and adults will be indoors only slightly more than they are now in the winter season. In the summer, however, having cooler weather will make it possible more often to be comfortable outdoors, beyond the confines of air-conditioned buildings. This is especially true in the southern half of the United States.

Neighboring.   By this term we refer to casual, informal chatting and visiting with those considered to be neighbors and nearby friends and relatives. Frequency of neighboring is known to be related to physical propinquity and ease of visual contact. This is especially true during those periods when friendships are forming.[10]

Under C and W conditions there will be less time spent outdoors and thus less opportunity for neighboring. This applies more to suburban residents with their yards, than to apartment dwellers and those living in the sparsely settled open countryside. We can expect fewer "spontaneous" outdoor barbecue picnics and less convivial beer drinking. It will take new residents in a neighborhood longer to "feel at home." Once friendships are established, there will be increased interaction on the telephone during the extended periods of inclement weather.

The C and D weather should see only moderate shifts in activity of the type noted above in the northern states but an increase in outdoor neighboring activity in the south where fewer very hot and humid days will make outdoor activity more comfortable.

Journey to Work.   It is rain, ice, and snow, rather than lower temperatures *per se* that have a significant impact of the journey to work. Thus, C and D conditions are relatively unimportant for this activity.

The anticipated C and W conditions mean more winter storms. Snow and ice storms produce widespread tardiness among employees. There are, of course, those relatively rare days when most business operations just shut down because it is believed that most employees won't be able to get to work. Absences approach 100%. But there are many more days when business and governmental operations are functioning after a fashion, but where absenteeism, due to the weather, is still quite high. Such weather conditions also mean that those going to and from work, especially in private vehicles, will be spending more time on the road and less at home. These generalizations will apply to approximately the northern two-thirds of the country, excepting the Pacific coastal areas.

Shopping.   Unlike the journey to work, most shopping can be postponed for a day or two. Critically needed medicines and baby foods are among the exceptions. There will be a tendency to stock up on critical items. More time will be spent shopping per item purchased under adverse weather conditions.

Education.   Since C and W conditions mean more winter storms and winter storms often produce unscheduled school closings, there will be more unexpected holidays from school for the children. For employed mothers of younger school children this creates a special strain since arranging for a daytime babysitter on short notice is difficult. Since the proportion of mothers in the labor force has been moving upward for several decades, this represents much more than a minor inconvenience. It represents an added threat to the health of children in such homes because of the increased likelihood of fire and accidents occurring in the absence of an adult in the house.

More cold, wet days also mean less outdoor exercise for children. When the weather is too foul the school play area stands empty as do the backyards of homes and the neighborhood parks. More children will get less active exercise in the winter months.

They will also have less social interaction with their playmates during such months. Mother can stand the noise of a dozen or more children in the backyard but only a few inside the house for very long.

Leisure and Recreation.   The term "leisure and recreation," as used here, refers to those activities apart from neighboring and family interaction that are valued as ends rather than as means to some other end. It includes an extremely wide range of activities from reading and bird watching to cross-country skiing and bowling.

Outdoor activities are the most obviously affected, but storms that make travel difficult and cause telephone and electric power outages can play havoc with indoor leisure and recreation also.

The C and D change will mean a reduction in snow-related sports activities. The investments in marginal ski areas where the snowfall is barely adequate now will be lost. On the other hand, for the few people who enjoy ice skating the situation should be improved on the local lakes, ponds, and rivers. With that exception, plus hiking, some hunting, fishing, and driving around to see the sights (assuming the fuel shortage doesn't last forever), winter will be a time of further reduced outdoor recreation and leisure.

Many of the spring, summer, and autumn sports should be enhanced by the lessened moisture. It is much easier to dress against the cold than against precipitation.

Should the change be to C and W, however, the situation will be more sharply changed from the status quo. While there is a growing recreation industry based on winter precipitation (snow), there is not much that Americans do with cooler but still liquid precipitation. True, added rainfall may contribute some to boating, water skiing, and, in appropriate amounts, to sport fishing, but there are not many other recreational activities that are enhanced by it. While there may still be a few lovers who like to go "walkin' in the rain," most people prefer to stay dry except, of course, when they are swimming, and even then they don't want rain. Cool, damp weather is bad for going to the beach, sunbathing, golfing, tennis, casual walking, baseball, and playing in the backyard with the children.

If the proportion of time spent at leisure and recreation remains constant, then it seems most likely that increasingly it will be done indoors with C and W conditions. We will probably see more time spent watching television and other passive, inactive forms of leisure. This will be especially true if cable television becomes a widespread phenomenon. Since children are less inclined to be physically passive in their leisure time than are adults, the changed climatic regime is likely to see parents more frequently playing family taxi driver as children are taken to the school gym, bowling alley, YWCA and YMCA, or anywhere else where something active is happening. Whether newspapers, magazines, and books will receive more attention is difficult to estimate. Television as a medium seems to still have a vast untapped potential to hold the attention of people of all ages.

Health Care.   Cooler and wetter conditions would seem to portend more illness and injuries. In addition to a probable increase in upper respiratory infections, we would expect more muscle strains and coronary problems from shoveling snow and pushing cars due to the increase in the number of snow storms. Snow and ice storms tend to produce more auto and pedestrian injuries and more falls due to slippery walking surfaces. Thus, there will be more time spent in home health care and at least some increase in the number of hospital patient days.

Political Activity.   Under C and W conditions there will be less citizen participation in the political process. As the proportion of really cold and wet days increases, the higher the probability that any given election day will have really nasty weather and thus a lowered turnout at the polls. Similarly, citizen participation in such activities as door-to-door canvassing and at outdoor political rallies will be lessened. We would expect, then, that there will be more efforts at organizing indoor political rallies and in using radio and television to get political messages across. It may be inevitable, anyhow, but the outcome over the years is likely to be that C and W conditions will be seen as contributing to lowered citizen participation prior to election day, and that trend will be used as a rationale to justify additional reliance on television to get a consideration of "the issues" before the public—shades of 1984!

Other Business Activity.   Detailed discussion of probable changes by type of business activity appears elsewhere in this volume. Here we note only a few general trends that may reasonably be expected.

Under C and W conditions there will be greater day-to-day variance in volume of retail sales. On "decent" days sales clerks will be rushed, while on "nasty" days the demand will be significantly less. During the colder months when snow and ice are a factor in transportation there will be more frequent weather-caused delays in securing inventory. In a similar vein, if there is really significant development of some form of "shopping by computer" then there will be additional days during the winter when the delivery of the selected retail items will be delayed due to hazardous driving conditions.

It is known that when there is a severe ice or snow storm many retail businesses experience sharp reductions in sales. It is not clear, however, whether those reductions simply represent postponed buying or whether some part represents lost sales. To the extent that for certain customers there is some impulse buying each time the person is in the store, then to the extent that storms reduce the number of times a person goes shopping during a year, they actually produce some reduction in total yearly retail sales. The larger the number of stormy days the greater will be the absolute reduction in sales.

The impacts discussed above should be markedly less under C and D conditions. Only to the extent that the total number of extremely cold days increases significantly would a measurable impact take place. Some reduction in severe winter storms would be anticipated under drier conditions, and this change would partially offset the negative impacts of the greater number of really extremely cold days.

Religious Activity. As the number of colder and wet days during spring, summer, and early fall increases, there will be some reduction, probably slight, in attendance at religious services. However, as the number of severe winter storms increases there will be sharp reductions in any religious activity that requires travel.

This section has treated probable changes in family and community activity. The emphasis has been on what will change within the community social system. In closing this section we discuss briefly what changes might be anticipated when the activity requires traveling to locations distant from the home community.

Impact of Curtailed Long Distance Travel

Curtailment of long distance travel is a winter phenomenon. Reduced precipitation means fewer snow and ice storm interruptions for ground and air transportation. But C and W conditions will increase the disruptions significantly.

American families are a mobile lot. Going to visit adult brothers, sisters, and parents usually means going a lot further than just across the city or half a country. The more affluent use commercial air carriers to get to long distance family gatherings. More winter storms mean more interrupted flights. But clearing an airport to resume flight schedules takes a lot less time than clearing the snow or ice from 500 miles or so of highways. Thus, with C and W conditions those who must travel by car will increasingly have more disruption and more canceled trips to family gatherings.

The same applies to long distance wintertime leisure and recreation trips. People committed to a few days or a week of skiing are perhaps less likely to cancel out due to severe weather. After all, snow is their thing. But people who would willingly drive 50 to 200 mi to a football game or a basketball tournament in decent weather will, in large numbers, stay home rather than risk accident on snow clogged or icy roads. And in the summer, camping trips and beach area vacations will be altered significantly by more cool, rainy, and overcast days.

With the possible exception of ski areas, those tourist communities now located in the humid areas of the United States will see a decline in income from customers normally coming from 50 or more miles away. This will apply to both winter and summer seasons, but more to the latter.

As the United States has become more and more suburbanized since World War II, there has been an increase in the proportion of workers who commute a "long distance" (25 or more miles one way) to work. With an increase in the number of days annually during which there are hazardous driving conditions, it is likely that long distance commuting to work will decline at least for those using cars.

We turn now to a consideration of the perception of climatic change by the nonexpert.

Perception of Climatic Change

We assume that whatever the climatic change produced by supersonic transport, it will develop very gradually. We also assume that those who have a vested interest in such transport will (1) be inclined to interpret the data in such a manner so as to conclude that no significant climatic change has or is likely to occur, and (2) emphasize the positive aspects of climatic change attributable to supersonic transport while playing down the negative consequences. The latter will occur primarily as a defense strategy after opponents of supersonic transport develop and present credible evidence on the negative impact of climatic change attributable to the aircraft.

There have been a number of studies of social aspects of planned weather modification.[7, 8] Here are some of the findings which may have a parallel with man-produced climatic change.

1. On the whole, people living in the area are not very well informed about intentional weather modification.
2. Persistent efforts to inform the public about an intentional weather modification program may gradually produce a more informed public.
3. Consistently, across areas, a sizeable minority believe that planned efforts at weather modification are very likely to upset the balance of nature.
4. The vast majority in every community studied hold that all decisions regarding possible weather modification projects should be made at the local level by the persons who may be affected by the results of the program.
5. Of those who actively oppose a weather modification program, the principal expressed reason is that the weather change has or will bring direct economic loss to them.
6. Severe, damaging weather events may provide a catalyst for organizing opposition to a weather modification project.
7. Citizens have little information about any relevant, long-term parameters of the climate in their area.
8. Where a concern for the preservation of the natural environment already exists, active opposition to intentional weather modification is more likely and will be based on environmental considerations.

Given the natural variability in the weather (some winters are much more extreme than those immediately preceding it), and given the lack of citizen knowledge about climatic data for their locale, there is every reason to believe that most citizens will not perceive any slowly developing climatic change.

Without such a perception they will not be inquiring about the possible causes of such change. And they are even less likely to organize in opposition to supersonic air transport.

But the generalizations just presented ignore one very important possibility, which is likely to change the general citizen perception and response significantly. That possibility is that highly organized and extremely competent environmental interest groups and organizations are very likely to play an important role in changing the perceptions of many citizens. Such organizations have a history of "success" in opposition to supersonic transport, and they are most unlikely to conveniently ignore the use of such aircraft after having invested so much in opposing their development. Furthermore, since climatic data are routinely gathered and published by the National Oceanic and Atmospheric Administration, only two points need to be established to make a convincing case for the public—that the climate has indeed changed and that the alteration has had significant negative effects. Data relevant to both points will not be too difficult to assemble.

Unless there is a demise of most environmental organizations within 5 to 10 years, it is a safe bet that such organizations will be instrumental, primarily through the mass media, in shaping the perceptions of a significant proportion of Americans regarding the negative consequences of supersonic transport-induced climatic change. Whether or not this emerging citizen concern will have much impact on the political process within the United States and on foreign relations between the United States and the countries sponsoring supersonic transport, are matters to be treated elsewhere. But it does seem safe to forecast that there will be persistent expression of citizen concern and that expression will be in opposition to the use of supersonic aircraft.

References

1. Adlam, T. N., *Snow Melting,* New York: The Industrial Press, 1950.
2. Brinkmann, Waltraud A. R., et al., *The Hurricane Hazard in the United States: A Research Assessment,* Boulder, Colorado: University of Colorado Institute of Behavioral Science, 1975.
3. Brinkmann, Waltraud A. R., et al., "Hail," *The Severe Local Storm Hazard in the United States: A Research Assessment,* Boulder, Colorado: University of Colorado Institute of Behavioral Science, 1975.
4. Brinkmann, Waltraud A. R., et al., "Tornado," *The Severe Local Storm Hazard in the United States: A Research Assessment,* Boulder, Colorado: University of Colorado Institute of Behavioral Science, 1975.
5. Cochrane, Harold, et al., *The Urban Snow Hazard in the United States: A Research Assessment,* Boulder, Colorado: University of Colorado Institute of Behavioral Science, 1975.

6. Grobecker, A. J., et al. *The Effects of Stratospheric Pollution by Aircraft,* Final Report. Cambridge, Mass.: Department of Transportation, Climatic Impact Assessment Program, December, 1974.

7. Haas, J. E., "Social Aspects of Weather Modification," *Bulletin of the American Meteorological Society,* Vol. 54, No. 7 (1973), pp. 647–657.

8. Haas, J. E., Boggs, K. S., and Bonner, E. J., "Science, Technology and the Public: The Case of Planned Weather Modification," *Social Behavior, Natural Resources and the Environment,* William Burch, et al., Eds. New York: Harper and Row, 1972.

9. Huszar, Paul C., *The Frost Hazard in the United States: A Research Assessment,* Boulder, Colorado: University of Colorado Institute of Behavioral Science, 1975.

10. Newcomb, T. M., *The Acquaintance Process,* New York: Holt, Rinehart and Winston, 1961.

11. Office of Emergency Preparedness, Executive Office of the President, *Report to Congress: Disaster Preparedness,* Vols. 1 and 3, Washington, D.C.: U.S. Government Printing Office, 1972.

12. Rogers, E. and Shoemaker, F., *Communication of Innovations,* New York: Free Press, 1971.

13. U.S. Department of Commerce, Environmental Science Services Administration, "Some Outstanding Snow Storms," #L. S. 6211, Washington, D.C.: U.S. Department of Commerce, 1966.

14. U.S. Department of Commerce, National Oceanic and Atmospheric Administration, *Climatological Data: National Summary, Annual 1970,* Ashville, N.C.: U.S. Department of Commerce, 1970.

15. White, Gilbert F., et al., *The Flood Hazard in the United States: A Research Assessment,* Boulder, Colorado: University of Colorado Institute of Behavioral Science, 1975.

16. White, Gilbert F. and Haas, J. Eugene, *Assessment of Research on Natural Hazards,* Cambridge, Mass.: MIT Press, 1975.

17. Wolford, V. L., *Tornado Occurrences in the United States,* Washington, D.C.: U.S. Weather Bureau, U.S. Department of Commerce, 1960.

5

Climate and Energy Demand: Fossil Fuels

JON P. NELSON
Department of Economics
The Pennsylvania State University
University Park, Pennsylvania

This study examines the influence of climate on the demand for fossil fuels in residential and commercial space heating. Climatic variables are represented by the number of heating degree-days, a measure of the duration and intensity of winter coldness. The demand for fossil fuels is taken to be the total oil, natural gas, and coal consumed in the residential and commercial sector in a cross section of states for 1971.

Our objectives in this chapter are twofold: first, to determine the covariation between heating degree days and fossil fuel energy demand while holding constant other determinants of demand (income, prices, urbanization, etc.); and second, to illustrate how the resulting empirical relationship might be used to predict changes in energy demand due to natural or man-made perturbations in the climate.

The plan of the chapter is as follows. The first section develops a simple, two-equation model of the demand for fossil fuels in residential and commercial space heating. One equation represents the level-of-utilization decision of a household or commercial establishment; a second equation represents the decision to adopt fossil fuel space heating equipment. The two-equation model is solved for a reduced form equation in either total demand or per capita terms. The next section presents estimates of the reduced form using ordinary least-squares regression analysis and cross-sectional data for 1971. The resulting degree-day parameter differs substantially from heating engineering expectations, and the parameter estimate is interpreted in light of this finding. The third section discusses several possible applications of the empirical results and indicates the energy costs likely to arise from supersonic transport. The last section contains a brief summary of the findings.

A Long-Run Demand Model for Space Heat

Space heating is an example of household production. Individuals and commercial establishments purchase production factors, including fuel, as inputs into the production of heat. Factors are *not* desired in and of themselves but only as intermediate goods in heat production. Joint products are also possible since some heating systems may produce different levels of cleanliness, humidity, and so forth. In this context the derived demand for fossil fuels is a function of (1) real income, (2) the relative prices of all goods consumed by the household, and (3) the ratios of relative prices of all commodities (factors) used to produce space heat to the price of space heat.[1]

Although the household production function is a useful framework in which to view the general empirical problem studied here, data requirements for such a model go considerably beyond the scope of the present paper.[2] An alternative is to view the household or commercial establishment as engaged in a two-stage decision process. In the first stage a decision must be made to adopt a given type of heating equipment, for example, fossil fuel or electric heating. In the second stage a decision must be made to determine the level of utilization at which to operate the equipment. Both decisions may be affected by economic, climatic, and noneconomic variables.

In this chapter these two decisions are considered simultaneously. The conceptual framework employed here is, therefore, more explicitly long-run than is sometimes the case. Energy demand studies which employ time-series data[3-8] result in parameters that are short-run estimates. That is, the empirical estimates primarily reflect level-of-utilization decisions and not the decision to adopt any particular type of heating equipment or, in the case of gasoline, size of automobile. The data employed in the present study are cross-sectional and will reflect long-run tendencies.

Studies of the demand for energy in the residential and commercial sectors can be divided into short-run models and long-run models. The short-run demand for a particular fuel refers to the demand for the fuel when the stock of heating appliances is held constant. Short-run per-customer demand is appropriately examined using time-series data that controls for the prices of substitute fuels and heating appliances. Long-run demand for a particular fuel refers to demand variations in which both the stock of appliances and the usage of that stock are allowed to vary. Long-run per capita demand can be examined using cross-sectional data that include the prices of substitute fuels and heating appliances. However, omission of the prices of heating appliances will not cause specification bias if these prices do not vary substantially across states. We shall assume this is the case since obtaining cross-sectional price variation is difficult for most commodities (fuels being an important exception.)[6] Moreover, the difficulty of obtaining accurate information on stocks of heating appliances suggests that the number of customers for each type of heating will have to serve as a substitute measure.

Total residential and commercial demand per state for fossil fuel energy is assumed to be a function of the number of customers, the price of fossil fuel energy, income, and several noneconomic variables including climate. Prices of substitute and complementary goods such as insulation, clothing, and possibly electricity are generally ignored due to cross-sectional data limitations. The number of customers per state using a particular heating appliance is in turn assumed to be a function of population size, the price of fossil fuel energy, prices of substitute fuels, income, and several noneconomic variables including climate.

Summing over the three types of fossil fuels, we have for the ith state the quantity-purchased or level-of-utilization demand relationship given by[9, 10]

$$TQ_i = c_0 + c_1 C_i + c_2 Y_i + c_3 P_i + c_4 DD_i + c_5 U_i + e_1 \qquad (1)$$

and for the number of customers relationship

$$C_i = d_0 + d_1 N_i + d_2 Y_i + d_3 P_i + d_4 PE_i + d_5 DD_i + d_6 U_i + e_2 \qquad (2)$$

where small c and d denote slope coefficients, and

TQ_i = total consumption of fossil fuel energy in the ith state
C_i = number of residential and commercial customers
N_i = population size
Y_i = income per capita
P_i = average price of fossil fuel energy
PE_i = average price of electricity
DD_i = average heating degree days
U_i = percentage urban population
e_1, e_2 = disturbance terms

Per capita income is included in both equations. The quantity purchased should be positively related to income, but the number of customers might be positively related to income at low levels and negatively related at high levels. At high levels of income, some individuals may find electric heating, for example, a preferable alternative because of greater cleanliness. The price of space heating energy is included in both equations to reflect fuel price substitution effects in both usage and investment. Negative coefficients for fuel price coefficients would be expected in both relationships. Ignoring some possibilities for short-run substitution or complementary effects, the price of electricity appears only in the customer equation. Its coefficient is expected to be positive.

Climate, measured by the number of degree-days, is included in both relationships. Colder weather will lead to increased usage over the heating season. For the ath day, degree days equals $65°F - T_a$, where T_a is the mean daily outdoor temperature, provided $T_a < 65°F$. At colder winter temperatures, more degree-days accumulate, and more fuel is required to maintain a given inside temperature. In addition, electrically heated homes are characterized by lower capital costs and higher operating costs. Therefore, as heating requirements increase, the total relative cost of electric home heating becomes less attractive.[11] Positive coefficients are expected in both relationships.

Urbanization is a proxy variable for several influences. Because population density and urbanization should be positively related to the availability of substitutes for space heating energy, negative coefficients are expected in both relationships. Urbanization should be negatively related to the number of single-unit structures and their size. Since single-unit structures tend to consume more energy due to size and exposure, a negative coefficient is implied for urbanization in the quantity-purchased relationship. Because multiunit structures may be served by a common heating unit, urbanization should be negatively related to the number of customers as well.

Both total quantity purchased, TQ, and the number of customers, C, are endogenous to the demand system. Substituting for C in equation (1), we have the following reduced form relationship:

$$
\begin{aligned}
TQ_i = (c_0 + c_1 d_0) &+ c_1 d_1 N_i + (c_1 d_2 + c_2) Y_i \\
&+ (c_1 d_3 + c_3) P_i + c_1 d_4 PE_i \\
&+ (c_1 d_5 + c_4) DD_i + (c_1 d_6 + c_5) U_i \\
&+ (e_1 + c_1 e_2)
\end{aligned}
\tag{3}
$$

which is easily rewritten as

$$
\begin{aligned}
TQ_i = a_0 &+ a_1 N_i + a_2 Y_i + a_3 P_i + a_4 PE_i \\
&+ a_5 DD_i + a_6 U_i + e_3
\end{aligned}
\tag{4}
$$

where, for example, $a_0 = (c_0 + c_1 d_0)$, $a_1 = c_1 d_1$, and so on.

Elasticities of demand derived from the structural demand equations measure the direct response of demand to changes in the explanatory variables and may be referred to as direct elasticities. These elasticities cannot be identified from the reduced form equation. Elasticities of demand derived from the reduced form equation will be referred to as total elasticities. For example, an increase in price will lead to a decrease in quantity purchased and in the number of customers. Measures of the total long-run response of demand to changes in the explanatory variables are obtained from the reduced form estimates.

Finally, it can be argued that the price of fossil fuel is also endogenous to the demand system. Under second-degree price discrimination, average price will vary with utilization so that the theoretically correct variables are marginal prices.[12, 13] In the long run, both average and marginal prices are relevant since average prices will affect the decision to adopt a given heating system while marginal prices are relevant in determining the level of utilization of the heating stock. However, the estimated price elasticities that would be obtained with marginal price data can be obtained from average price data if the estimated demand equation is log-linear.[10] The energy prices employed in this study are *ex post* average prices derived from average revenue data.

Empirical Results

The reduced form model of the previous section was estimated using ordinary least squares regression analysis and cross-sectional data by state for the year 1971. The following variables and data sources were used:

TQ_i = total oil, gas, and coal consumed in the residential and commercial sector in the ith state in 1971. Measured in 10^{12} Btu.[14]

N_i = total population in the ith state in 1971. Measured in 10^3 people.[15]

Q_i = total oil, gas, and coal consumed per capita in the residential and commercial sector in the ith state in 1971. Measured in 10^6 Btu per capita.[14, 15]

Y_i = personal income per capita in the ith state in 1971. Measured in dollars per capita.[15]

P_i = weighted average residential and commercial price of oil, gas, and coal in the ith state in 1971. The weights are the relative proportions of each fuel consumed in 1971 by state and sector. Measured in dollars per 10^6 Btu.[16, 17, 18]

PE_i = average residential and commercial price of electricity in the ith state in 1971. Measured in dollars per 100 Kwh.[19]

DD_i = weighted average heating degree-days for the ith state in 1971. The weights are the relative 1970 populations of Standard Metropolitan Statistical Areas (SMSAs) and selected cities with reporting weather stations. From one to nine SMSAs and cities were used for each state. Measured in degree-days per year.[20]

U_i = percentage of total population in the ith state living in urban areas in 1970. Measured in percentage points.[15]

Equation (3) was estimated in both linear and log-linear forms. Application

Table 1. Regression Results with Total Fuel Consumption (TQ) as the Dependent Variable

Variable[a]	Regression		Beta Coefficient Regression 1
	1	2	
Constant	−4.4116	−3.4830	
	(0.583)*	(0.755)*	
N	0.9447	0.9427	0.991
	(0.024)*	(0.024)*	
Y	0.4581	0.0127	0.068
	(0.178)*	(0.296)	
P	−0.3250	−0.2311	−0.090
	(0.084)*	(0.096)*	
PE	0.5244	0.5308	0.114
	(0.103)*	(0.101)*	
DD	0.4503	0.4856	0.254
	(0.043)*	(0.046)*	
U		0.2980	
		(0.1604)**	
R^2	0.983	0.984	
$\bar{R}^2$	0.980	0.982	
F-ratio	473.144	417.900	
Multico[a]	0.422	0.109	

* Significant at the 95% confidence level, two-tailed t test. Estimated standard errors in parentheses. All variables in common logs, sample size $n = 48$.
** Significant at the 90% confidence level.
[a] Multico is the determinant of the correlation matrix of independent variables that equals one for perfect orthogonality and zero for perfect collinearity.

of a nonparametric statistical test suggested by Box and Cox[21, 22] indicated that the log-linear form provided a superior fit.[23] Table 1 presents the parameter estimates for equation (3) in which all variables are expressed in common logarithms. Consequently, the resulting coefficients are constant demand elasticity estimates for population, income, fuel price, electricity price, degree-days, and urbanization.

The results in Table 1 indicate the following:

1. Population has a significant and positive effect on total fuel demand. The elasticity is close to unity in value and indicates that the model could be recast in terms of per capita demand.[22, 24]

2. Income has a significant and positive effect on total fuel demand in regression 1. However, when the urbanization variable is added, the income variable is no longer significant. This result may be due to the high correlation ($r = 0.749$) between these two variables.

3. Total fuel demand is price inelastic for residential and commercial space heating.

4. The electricity price variable had a positive and significant effect on total fuel demand. The cross-price elasticity of demand is approximately 0.5.

5. Degree-days has a positive and significant effect on total fuel demand. However, the elasticity is less than one; a 10% increase in degree-days would only increase total fuel demand by about 5%.

6. The beta coefficients for regression 1 indicate that the two most important variables in explaining changes in total fuel demand are population and climate (degree days).

Table 2. Regression Results with Per Capita Fuel Consumption (Q) as the Dependent Variable

	Regression		Beta Coefficient
Variable[a]	1	2	Regression 1
Constant	−1.0771	−0.1920	
	(0.591)**	(0.788)	
Y	0.2671	−0.1584	0.111
	(0.165)	(0.304)	
P	−0.2800	−0.1901	−0.215
	(0.086)*	(0.100)**	
PE	0.5039	0.5092	0.305
	(0.108)*	(0.106)*	
DD	0.4955	0.5301	0.780
	(0.040)*	(0.045)*	
U		0.2804	
		(0.169)	
R^2	0.847	0.857	
$\bar{R}^2$	0.833	0.840	
F-ratio	59.696	50.243	
Multico[a]	0.657	0.170	

* Significant at the 95% confidence level, two-tailed t test. Estimated standard errors in parentheses. All variables in common logs, sample size $n = 48$.
** Significant at the 90% confidence level.
[a] Multico is the determinant of the correlation matrix of independent variables that equals one for perfect orthogonality and zero for perfect collinearity.

Since population exhibits approximately unit elasticity, the model was recast in terms of per capita demand for fossil fuel. The resulting relationship is, of course, closer to a theoretically plausible demand function since the dependent variable now refers to individual behavior, at least on the average for each state. These new results are presented in Table 2.

With per capita fuel consumption Q as the dependent variable, the results indicate the following:

1. Income is not significant in either regression and becomes negative when the urbanization variable is added. This result may be due in part to consumer preferences for electric heating at higher income levels.
2. Price of fossil fuels, the electricity price variable, and degree days all retain their significance and relative magnitudes.
3. The beta coefficients for regression 1 indicate that climate (degree-days) and the price of electricity are most important for per capita demand changes.

In evaluating the empirical results, it is significant that the degree-day variable is less than unity. In economic terms, the engineering production function for residential space heating is said to be homogeneous of degree one.[25, 26] Yet the results in Tables 1 and 2 indicate an elasticity of about 0.5 rather than 1.0. There are several possible explanations for this result[26]: (1) incorporation of an insulation effect in cross-sectional state data on degree days; (2) inclusion of a nonspace heating component in the dependent variable; and (3) aggregation of the residential and commercial sectors in the dependent variable.*†

First, the empirical model does not control for interstate differences in housing insulation and construction standards. We may assume that relatively higher prices for energy relative to construction and insulation materials and higher annual heating costs in general induce an insulation effect in housing construction. Because homes in northern climates are better insulated, the degree-day variable will be biased downward if estimated with cross-sectional data. For a 10% interstate increase in degree-days, energy consumption will increase by less than 10% if housing in northern climates contains greater amounts of insulation or generally sounder construction. This effect is

---

* As heating requirements increase, the total relative cost of electric home heating increases (Ref. 11, p. 16). The degree-day elasticity will contain a positive component reflecting the increased economic incentive to adopt fossil fuel space heating in colder climates.
† Because fossil fuel demand is a derived demand, its usage will depend on the elasticity of substitution between fuel and other factors (e.g., insulation) and on the elasticity of demand for the final commodity, indoor temperature, and comfort. To the extent that there is some elasticity in the demand for indoor temperature in a cross section of states, individuals in colder climates may reduce fuel usage by lowering their thermostats, shutting off unused rooms, and so on. As a consequence, the degree-day elasticity will be less than unity.

illustrated in the following statement with regard to natural gas consumption in home heating:

Although the relationship between estimated fuel consumption, design heat loss, design temperature difference and degree days is non-linear, factors for gas . . . as corrected if necessary, are satisfactory for regions having 3500 to 6500 degree days per heating season. In regions with less than 3500 degree-days, the unit gas consumption is *higher* than given; where over 6500, the unit is *less* than given. Ten percent addition or deduction in these cases is recommended by A.G.A. publications.[25]

Second, the dependent variables examined in this study included some non-space heating uses of fossil fuels. A rough estimate is that space heating accounts for 80% of the gas, oil, and coal consumed in the residential and commercial sector.[27] To the extent that the nonspace heating component does not vary with outdoor temperature, the degree day elasticity measure will be biased downward.

Finally, the engineering degree-day elasticity for commercial buildings must be determined on a case-by-case basis.[25] Consequently, it can only be said that the degree day elasticity derived in this study is a weighted average for the residential and commercial sector.

Applications

The continental United States contains a variety of climates measured by heating degree-days (Table 3). The 30-year normal for average degree-days per year ranges from 759 days in Florida to 9097 days in North Dakota. Fossil fuel energy consumption in the year 1971 varied from 20.8 million Btu per capita in Florida to 155.4 million Btu per capita in Wyoming. However, as was demonstrated in the previous section, the observed variation in per capita energy consumption is only partly due to climatic variation. This section presents two applications of the empirical results: first, an evaluation of the expected crude petroleum savings due to lower thermostat settings and higher prices for fuel oil; and second, a prediction of increased energy consumption *if* supersonic air transportation were to reduce mean annual global surface temperatures.

The number of heating degree-days per day is defined as the difference between 65°F and the mean daily outdoor temperature. The application of the degree-day concept in engineering estimates of space heating energy requirements assumes continuous heating to 68–72°F.[25] The difference between the 65°F base and the indoor temperature is due to food cooking, clothes drying, hot water pipes, electric lighting, people, and so forth. For a particular year, the number of heating degree-days is simply the total accumulated number of days for that period. Normal yearly heating degree-days is defined as the 30-

Table 3.   Normal Heating Degree-Days and Per Capita Fossil Fuel Energy Consumption in 1971, by State

Climatic Zone and State	Normal Degree Days per Year[a] (30-yr average)	1971 Fossil Fuel Energy Consumption (10^6 Btu/Capita)
0 to 3500 Normal Degree-Days		
Florida	759	20.8
Louisiana	1716	50.6
Arizona	1763	48.0
Texas	1979	46.8
California	2143	50.8
Alabama	2216	42.7
Mississippi	2216	50.4
South Carolina	2518	39.3
Georgia	2870	44.8
Arkansas	3231	69.3
Tennessee	3407	42.3
Nevada	3473	56.2
North Carolina	3484	46.5
3500 to 6500 Normal Degree-Days		
Virginia	3652	47.7
Oklahoma	3783	71.1
New Mexico	4294	66.4
West Virginia	4505	57.1
Maryland—D.C.	4569	71.4
Oregon	4675	55.7
Kentucky	4727	58.7
Kansas	4776	87.4
Missouri	4860	77.2
Delaware	4930	76.3
New Jersey	5050	87.7
New York	5360	81.0
Washington	5362	50.1
Pennsylvania	5491	71.4
Indiana	5794	88.6
Massachusetts	5846	114.4
Connecticut	5898	66.0
Ohio	5943	82.3
Rhode Island	5945	97.4
Utah	6052	87.1

Table 3. Continued

Climatic Zone and State	Normal Degree Days per Year[a] (30-yr average)	1971 Fossil Fuel Energy Consumption (10^6 Btu/Capita)
Idaho	6107	72.4
Illinois	6141	97.1
Nebraska	6147	93.5
Colorado	6217	91.2
Michigan	6423	84.0
Over 6500 Normal Degree-Days		
Iowa	6862	87.4
New Hampshire	7383	91.9
Wyoming	7417	155.3
Maine	7511	97.8
Montana	7677	97.5
Wisconsin	7689	85.9
South Dakota	7840	82.2
Vermont	8269	82.8
Minnesota	8470	87.6
North Dakota	9097	81.3

Source: References 14, 15, and 20.

[a] Weighted average computed using 1970 population weights for reporting weather stations.

year average for 1931 to 1960. For the ith weather station, an expression for normal yearly heating degree-days is given by

$$\overline{DD}_i = \sum_{a=1}^{A} (B - T_{ai})$$
$$= A(B - \bar{T}_i) \tag{5}$$

where

$\overline{DD}_i$ = normal total heating degree days per year for the ith weather station

B = reference base temperature, a constant which is equal to 65°F in most U.S. weather applications

T_{ai} = normal mean outdoor temperature on the ath day for the ith weather station, provided $T_{ai} < B$

A = normal number of days per year for which $T_{ai} < B$

T_i = normal mean outdoor temperature for the period $a = 1, \ldots, A$, $\bar{T}_i < B$

A change in average ambient indoor temperature due to a lowering of thermostats may be viewed as a change in the reference base from 65°F to some smaller base. A basic expression for determining the effect of a lowering of thermostats on degree-days, and hence on energy demand, is approximately

$$\Delta \overline{DD}_i = A \, \Delta B \qquad (6)$$

Application of equation (6) requires knowledge of the parameter A, the normal number of days per year for which the normal mean daily outdoor temperature is less than 65°F. To estimate A, data were obtained for 267 cities for which $\overline{T}_i < 65°F$.[25] Since it is known *a priori* that $B = 65°F$, a linear regression was performed for equation (5) with total normal heating degree-days per year per city as the dependent variable and $65 - \overline{T}_i$ per city as the independent variable. In this case, $\overline{T}_i$ was the normal mean daily outdoor temperature for the 212-day period of October 1 to April 30. The regression results using least squares were

$$\hat{\overline{DD}}_i = 230.33 \, (65 - \overline{T}_i); \qquad R^2 = 0.986 \qquad (7)$$
$$\underset{(0.667)}{}$$

with the standard error for A in parentheses. Regressions were also run for this relationship using the method of group averages; the estimate of A was 227.78 days. Separate regressions by least squares for groups of cities in three climatic zones (see Table 3) yielded estimates of A that ranged from 226 to 237 days.

A reduction in all residential and commercial thermostats by 6°F would indicate a reduction in degree-days by $\Delta DD = A \, \Delta B = 230 \, (-6) = -1380$ days per year. An average value for heating degree-days per year for the continental United States is 5053 days. Thus, the implied proportional reduction in normal degree-days is 1380/5053 = 0.273, or 27.3%. Given this value, the reduction in energy demand depends on the degree-day elasticity of demand. For an elasticity of 0.5, *per capita* demand would be reduced by 13.65%. If it is assumed that these calculations can be applied to fuel oil demand alone, a 6°F reduction in thermostat settings and an elasticity of 0.5 imply an average daily savings of 420,000 barrels of crude petroleum, or about 12% of residential and commercial petroleum demand in 1973, everything else constant.[28]

Price increases for fuel oil will also have a conserving effect on petroleum demand, especially in the long run. For a price increase of 10 cents per gallon relative to November–December 1973 prices and a price elasticity of demand of −0.3, the savings in barrels of crude oil per day would be approximately 275,000 barrels or 8% of 1973 demand, everything else constant.[28]

Cyclical, secular, or man-made perturbations in the climate will also influence energy demand. Propulsion emission by supersonic aircraft in the stratosphere is one possible example of man-induced climatic change.[29] For example, a $-1°C$ ($-1.8°F$) change in the mean annual global temperature implies an increase in degree-days of $\Delta DD = A \Delta T = 230 (1.8) = 414$ days per year. The proportional change in degree-days is $414/5053 = 0.082$, or 8.2%.

A permanent decrease in mean annual global temperatures implies a stream of incremental costs from the present to perpetuity. The size of the cost stream will depend on total fossil fuel demand, which in turn will reflect population, income, and relative energy prices. In developing predictions for global temperature changes, it was assumed that the population of the United States in the year 2025 would be 284 million and that real personal income per capita would be \$11,480 (in constant 1971 dollars). Relative prices for fuels and other commodities were assumed to be constant.[29] Forecasts for per capita fuel demand were prepared for each year under constant climatic conditions. Incremental demand due to decreased global temperatures then depends on the degree-day elasticity of demand. For an elasticity of 0.5, a decrease in the mean annual global temperature by $1.8°F$ implies a 4.1% increase in per capita fuel demand per year, everything else constant. That is, an 8.2% increase in degree-days in the United States due to climatic change will increase per capita residential and commercial demand for fossil fuels by 4.1%. Alternatively, an elasticity of 1.0 implies an 8.2% increase in demand.

The stream of added costs due to a mean annual global temperature change of $-1.0°C$ and a degree-day elasticity of 0.5 was discounted to 1974 using interest rates of 3, 5, or 8%. For an interest rate of 5%, the present value of added fossil fuel costs would be \$3516 million. On an annual basis this is equivalent to \$175.8 million per year added energy costs.[29] As a proportion of Gross National Product (GNP), an annual cost of \$175.8 million is a rather small sum. A complete analysis of the costs and benefits of climatic modification would, of course, incorporate a number of possible effects, both quantifiable and qualitative.* In addition, the present value of added fuel costs is a much larger sum, and much of this burden, like the public debt, will be borne by future generations.

Summary

The recent energy crisis has stimulated considerable interest in econometric modeling of energy demand and supply. From an economist's viewpoint, the

---

* A more complete benefit-cost analysis of climatic modification is found in reference 29. The computations presented here reflect the movement along a Btu-degree-day relationship or a shift in the demand schedule due to climatic modification. Since the increase in expenditures resulting from the demand shift is avoidable, it is a cost of climatic change. In addition, given a supply schedule that is less than perfectly elastic, there will be a loss of consumer surplus resulting from a higher price relative to the old demand schedule. This cost is not accounted for in the computations.

main demand variables of interest are relative energy prices and income. Climate, when taken explicitly into account, is usually found to be important in explaining the demand for space heating and air conditioners.[8, 10, 11] In the present study, climate (heating degree-days) was the most important explanatory variable in explaining per capita fossil fuel (oil, gas, and coal) demand and was second only to population in explaining total fossil fuel demand. Clearly, future econometric energy analysis must carefully consider the effects of climate prior to the incorporation of secondary explanatory variables. Future research should also be directed to regional, seasonal, and daily variations in climate and energy demand. Although climate surely is not a policy variable of the magnitude of energy prices or income, man-made or secular changes in climate are possible. Studies such as the present one can add to our scarce knowledge[30] of the value of the weather and how it is associated with space heating and with the heating industry.

References

1. Muth, R. F., *Econometrica,* Vol. 34 (1966), p. 703.
2. Blair, L. M., "Household Production Functions as a Basis for Derived Demand Functions: The Case of Automobiles," unpublished paper, University of Utah, Salt Lake City, no date.
3. Balestra, P., *The Demand for Natural Gas in the United States,* Amsterdam: North-Holland, 1967.
4. Erickson, E. W., Spann, R. M., and Ciliano, R., "Substitution and Usage in Energy Demand: An Econometric Estimation of Long-Run and Short-Run Effects," in M. F. Searl, Ed., *Energy Modeling,* Washington, D.C.: Resources for the Future, Inc., March 1973, pp. 190–208.
5. Fisher, F. M., and Kaysen, C., *The Demand for Electricity in the United States,* Amsterdam: North-Holland, 1962.
6. Houthakker, H. S., and Taylor, L. D., *Consumer Demand in the United States,* 2nd ed., Cambridge: Harvard University Press, 1970, p. 277.
7. Houthakker, H. S., Verleger, P. K., and Sheehan, D. P., *American Journal of Agricultural Economics,* Vol. 56 (1974), p. 412.
8. Stout, A. M., *Review of Economics and Statistics,* Vol. 43 (1961), p. 185.
9. Anderson, K. P., *Journal of Business,* Vol. 46 (1973), p. 526.
10. Halvorsen, R., *Review of Economics and Statistics,* Vol. 57 (1975), p. 2.
11. Wilson, J. W., *Quarterly Review of Economics and Business,* Vol. 11 (1971), p. 16.
12. Houthakker, H. S., *Journal of the Royal Statistical Association (A),* Vol. 114 (1951), p. 351.
13. Taylor, L. D., *Bell Journal of Economics and Management Science,* Vol. 6 (1975), p. 74.

14. U.S. Department of Interior, *United States Energy Fact Sheets by States and Regions, 1971,* Washington, D.C.: U.S. Department of Interior, February 1973.

15. U.S. Department of Commerce, *Statistical Abstract of the United States, 1971,* Washington, D.C.: U.S. Government Printing Office, 1971.

16. American Gas Association, *Gas Facts: 1971 Data,* Arlington: American Gas Association, 1972.

17. "Fueloil Markets," *Fueloil and Oil Heat,* Vol. 3 (December 1971), p. 22.

18. U.S. Department of Labor, *Retail Prices and Indexes of Fuels and Utilities,* Washington, D.C.: U.S. Bureau of Labor Statistics, November and December 1971.

19. Edison Electric Institute, *Statistical Yearbook of the Electric Utility Industry for 1971,* New York: Edison Electric Institute, 1971.

20. U.S. Department of Commerce, *Climatological Data: National Summary for 1971,* Washington, D.C.: U.S. Government Printing Office, 1971.

21. Box, G. E., and Cox, D. R., *Journal of the Royal Statistical Society (B),* Vol. 26 (1964), p. 211.

22. Rao, P., and Miller, R. L., *Applied Econometrics,* Belmont: Wadsworth, 1971, pp. 40–43, 107–111.

23. Nelson, J. P., *Review of Economics and Statistics,* Vol. 57 (1975), p. 508.

24. Mount, T. D., Chapman, L. D., and Tyrrell, T. J., "Electricity Demand in the United States: An Econometric Analysis," Oak Ridge, Tenn.: Oak Ridge National Laboratory, June 1973.

25. ASHRAE, *Guide and Data Book: Systems, 1970,* New York: American Society of Heating, Refrigerating and Air Conditioning Engineers, 1970, pp. 621, 626. 628, 632–633.

26. Nelson, J. P., "Engineering Production Functions and Climatic Modifications," in Reference 29, pp. 165–185.

27. Stanford Research Institute and U.S. Office of Science and Technology, *Patterns of Energy Consumption in the United States,* Washington, D.C.: U.S. Government Printing Office, January 1972.

28. Ferrar, T. A., and Nelson, J. P., *Science,* Vol. 187 (21 February 1975), p. 644.

29. U.S. Department of Transportation, Climatic Impact Assessment Program, *Economic and Social Measures of Biologic and Climatic Change,* CIAP Monograph 6 (DOT-TST-75-56), National Technical Information Service, Springfield, Virginia, September 1975.

30. Maunder, W. J., *The Value of the Weather,* London: Methuen, 1970.

6

Electricity Demand in All-Electric Commercial Buildings: The Effect of Climate

THOMAS D. CROCKER
Department of Economics
University of Wyoming
Laramie, Wyoming

Since the advent of the "energy crisis," a number of scholarly papers have attempted to estimate the demand for electricity. These papers extend either in concept, empirical technique, and/or data set employed the earlier published efforts of Fisher and Kaysen,[6] Houthakker,[11] and Wilson.[20] In general, they are in the tradition of this earlier work in that they use highly aggregated data, are devoted either to the residential or manufacturing sectors, and give only cursory attention to the covariation of climate and electricity consumption since they disregard the use of electricity for space heating and cooling. The objective of this chapter is to remedy a combination of these omissions; that is, completely disaggregated data is employed to estimate for a sample of all-electric commercial buildings the covariation of electricity consumption and several measures of climate.

The assistance of Larry Eubanks and Robert Horst, Jr. is gratefully acknowledged. This research was supported by the Climatic Impact Assessment Program of the U.S. Department of Transportation.

The possible biases introduced by using aggregated data, such as the U.S. statewide measures employed in many studies of electricity demand, are well explained by Green[7] and need not be detailed here.

According to Chapman[4], the total electricity per family used by the commercial (nonresidential and nonindustrial) sector in the United States in 1968 was 5232 kWh. This represented in 1968 a not insignificant 24% of the total electricity per family for all uses. Yet nowhere in the literature, to the best of this writer's knowledge, is there a study of the commercial demand for electricity. Usually, the commercial sector is lumped with the residential sector or is altogether disregarded. If estimates of the responsiveness of electricity consumption to changes in some parameters are to be used for the demand projections essential to planning future energy production, it is perhaps important that separate and distinct estimates be made for the residential sector and the commercial sector. Moreover, the sheer bulk of many commercial buildings makes economical the use of certain space-heating technologies such as high-intensity lighting that are impractical in buildings used for residential purposes.

Most studies of the demand for electricity that have devoted any attention whatsoever to climatic factors have focused upon the residential sector. Generally, as in Anderson,[1] these studies have indicated that climatic variations contribute little if anything to explaining variations in the quantity of electricity demanded. In fact, whatever influence the estimation procedures imply climate has upon electricity demand can, by commonly accepted tests of statistical significance, be attributed to chance rather than to a relation between climate and electricity demand. These results are perhaps not terribly offensive to one's intuition if one believes most electricity is consumed in the provision of illumination and the operation of machinery, including home appliances. In fact, according to Chapman,[4] 5348 kWh of the 6743 kWh used for residential purposes by the average U.S. family in 1968 was devoted to illumination and machinery operation. One could reasonably expect, at least as a first approximation for both the residential and commercial sectors, that consumption of electricity for these purposes is more or less insensitive to climatic variations within the United States. However, the result becomes rather less believable with an awareness that the unit of analysis employed in these studies of the residential demand for electricity is typically an arithmetic mean building unit, where each mean is meant to apply to a single state. If the population of building units for which the mean is calculated includes a fair number in which heating and/or cooling is directly provided by electricity, the aforementioned results seem quite questionable. They seem even more questionable given that approximately 90% of all building space-cooling equipment in 1970 was powered by electricity,[5] that electric power consumption as perceived by electricity producers ". . . is highly related to daily temperatures in distribution areas,"[14] that electricity used for space heating increased from 4% of total electricity consumption in all sectors in 1950 to 16% in 1970,[16] that the fraction of new homes with electric space heating increased from 20% in 1966 to 28% in 1968,[16] and that builder concern about the availability of oil and gas is purported to have caused the number of all-electric homes built to have doubled in Winter 1973–1974.[2] Furthermore, Anderson[1] states that in the all-electric home, electric central heating ". . . accounts for one-half to three-fourths of total electricity used." Of the data for individual commercial buildings used in the empirical portion of the present study, 50 to 70% of the total connected load in each building unit is, according to the engineering specifications, dedicated to space heating and cooling. Thus, although much of the aforementioned support for the hypothesis of a nonzero responsiveness of electricity consumption to climatic variations is drawn from the residential sector, it is by no means implausible that the sign of the measure of response would be the same in the residential and commercial sectors, although the magnitude of the measure might well differ between the two sectors.

The fundamental purpose of this chapter is to estimate, using completely disaggregated data for commercial structures, the covariation of electricity demand and climatic indicators. The data employed are detailed histories of actual, month-by-month electricity consumption and weather environment over a period of a year for about 80 all-electric commercial buildings throughout the United States.*

Some Remarks on a Theoretical Framework

The demand for electricity and other energy forms for space heating and cooling is a derived demand. That is, electricity in and of itself does not yield utility or revenues. It does so only by serving as an input which, when combined with other inputs, produces an output that does, in fact, yield utility or generate revenues. The demand for the input, electricity, is thus derived from the demand for the output in question, the building service represented by the temperature for the volume of air enclosed by a building unit. Given a change in exterior climate, the air temperature inside building units may be maintained by altering the consumption of electricity, substituting another form of energy, changing the extent of use of some building attribute such as insulation, or some combination of all three. With a change in a variable he is unable to manipulate, for example, climate or the electricity price schedule, the decision maker will respond in the short run by altering his consumption of electricity. However, in the long run, the decision maker will be able to adjust the

* These case histories were collected by the Electric Heating Association, Inc., 437 Madison Avenue, New York 10022. I am grateful to Mr. Jack Day of the Commonwealth Edison Company, Chicago, Illinois, who graciously provided the case histories and generously took the time to aid the author in their interpretation.

mix and magnitude of building attributes he employs and the energy forms he uses either by relocating or by constructing new building units. Implicit in this explanation is a "putty-clay" view of the space heating and cooling equipment that will be consumptive of electricity. The investment possibilities are like putty. However, once the investment decisions have been made and actually completed, the space-heating and cooling equipment already in place is not costlessly removed. The putty has hardened into clay.

Climatic variables are not viewed as direct factors of production, subject to manipulation by the decision maker. Instead, these variables operate through changes in the production function relating building inputs to the interior air temperature of the building. These climatic variables thus affect the interior air temperature by changing the productivities of the direct factors of production or by affecting the choice of the production process employed. Changes in climate can therefore be viewed as analogous to technological changes.

It is well known from the work of Hicks[10] and others that among the determinants of the elasticity of derived demand for a factor of production in a particular technology are the elasticity of demand for the final product, the elasticity of supply of other factors of production, the elasticity of substitution between one factor of production and another, and the relative importance of the factor in the production process. In general, the greater is the absolute magnitude of any one of these measures, the greater will be the elasticity of derived demand for any input factor. However, an important exception occurs when the elasticity of demand for the output fails to exceed the elasticity of substitution of the factor in production. In this case, the elasticity of derived demand for the factor will not vary directly with the relative importance of the factor input.

Nearly all empirical studies of the demand for electricity do not permit one to account for the influence of the above measures upon the elasticity of derived demand of electricity used for space-heating and cooling purposes.* The present study is no exception. That is, these studies, including the present one, do not permit the decision between electricity consumption and the mix and magnitude of the stock of building attributes and energy forms to be simultaneously determined. The effects of changes in variables that the decision maker is unable to manipulate upon the decision maker's choice between electricity consumption and other factor inputs are therefore neglected. For example, the influence of climate upon such measures as the elasticity of substitution between electricity usage and usage of building materials, and, through this measure, the ultimate effects upon the elasticity of derived demand for electricity used for space heating and cooling are disregarded.

Usually, the aforementioned studies develop a single-equation model and

* In addition to those already mentioned, among the foremost of these studies are Griffin,[8] Halvorsen,[9] and Houthakker and Sheehan[12] in the residential sector, and Baxter and Rees[9] and Griffin[8] in the industrial sector.

then estimate its parameters. For example, in the present study, the quantity of electricity consumed by month in a sample of individual commercial, all-electric buildings is related to measures of local climate and other variables, including variables intended to represent those structural attributes of each building that can be substituted for electricity consumption. The values of the estimated parameters for climate thus show the variation in electricity consumption to be expected, on average, for a one-unit change in the climate variables. However, underlying the estimates obtained is the assumption that the structural attributes of each building are invariant. That is, one can use the estimated parameter to calculate the affect of a climate change upon electricity consumption if and only if one is willing to assume that building attributes will not also be altered in response to the hypothesized climate change. To the extent, then, that the climate elasticity of derived demand for space heating and cooling does not exceed the elasticity of substitution of electricity for construction materials and other energy forms, the absolute magnitudes of the climate parameters of most empirical studies of the demand for electricity, including the present one, are biased downward.

The alleviation of these problems requires the construction and estimation of a model permitting the simultaneous determination of the demand for alternative energy forms, building materials, and electricity. Of the models thus far published, only Griffin's[8] approaches a treatment of this sort. However, if, for empirical purposes, it is assumed that climatic changes are technologically "factor neutral" in a Hicksian sense, the problem of accounting for changes in the elasticity of derived demand for electricity caused, for example, by changes in the elasticity of substitution of electricity for building materials can be avoided. The reason is that under this assumption, climate affects the marginal products of the factor inputs proportionately. That is, the marginal rate at which electricity is substituted for other inputs is unaltered. If, in addition, it is assumed that climatic changes exert no influence on the climate elasticity of demand for building services and the elasticity of supply for other factor inputs, then one can dismiss the simultaneity problem.

There is no a priori reason to expect the technology of electricity consumption to conform to the above structures. These are simply the assumptions that must be made if the estimates in this chapter of the responsiveness of commercial electricity consumption to changes in climate are to be taken as conclusive. The available data did not permit estimation of the other market parameters faced by the decision makers responsible for planning and constructing the buildings of the sample. Since explicit consideration of these parameters would tend to expand the number of substitution possibilities said to have been available to decision makers, the estimates presented here are best viewed as lower bounds of the long-run climate elasticities of demand in all-electric, commercial buildings.

The estimates of this chapter are considered to be long-run because the sample of buildings is cross-sectional. In effect, this implies that factor inputs such as building insulation and heating and cooling equipment are selected on the basis of expected climate, where expected climate may include a range of possible short-term variations in weather from one year to another that do not make the transition to other combinations of factor inputs worthwhile to the decision maker.* In the stochastic form of the derived-demand function to be estimated, a multiplicative random error term is used in the conventional way, assuming, in particular, that this error is distributed independently of the levels of all environmental influences, outputs, inputs, and input prices. Thus if the random variate is taken, for example as a measure of unexpected climatic variations, we may assume that inputs, such as electricity and building materials, are chosen on the basis of an expected climate that allows for some weather variability from year to year. The amount of electricity consumed is therefore independent of the particular realization of the random variate.

The empirical estimates obtained of the derived demand for electricity assume a Cobb-Douglas production function with no restriction in the parameters for the transformation of electricity and building materials into interior building air temperatures. This implies that the expression to be estimated will be multiplicative, that is, linear in the logarithms of the original variables. There exists no compelling reason for the selection of this functional form other than that it has been found to be serviceable in other studies of electricity demand, for example, Baxter and Rees,[3] Houthakker,[11] and Anderson.[1]

The Data

As noted earlier, the data employed in this study are detailed histories of actual, month-by-month electricity consumption and weather environments over a year for 81 all-electric, commercial buildings throughout the United States. The year for which the history of each building was recorded was some 12-month period in the interval 1963 to 1971.

Table 1 is a summary of the meanings, the units of measure, and the expected signs for variables to be used in the empirical analysis. The dependent variable in each of the ordinary-least-squares, cross-sectional regressions is kilowatt-hours consumed in a given month. Since the data did not permit dis-

---

* Rather than assuming that the long-run equilibrium choice of building materials and energy forms chosen by the decision maker takes into account year-to-year variability in local weather, one might suppose, in accordance with Fisher and Kaysen[6] and Houthakker and Taylor,[13] that changes in climate and other exogenous parameters lead to discrepancies between existing and desired combinations of building materials and electricity usage. The discrepancy then sets into motion a time-distributed effort on the decision maker's part to revise the existing combination. His rate of revision is dependent on the magnitude of the discrepancy, relative prices, and so forth.

Table 1.   Definitions of Variables and Their Expected Signs in the Regressions[9]

Variable Symbol	Definition	Unit of Measure	Expected Sign
kWh	Kilowatt-hours	Monthly	Dependent
DD	Heating degrees	Days/month	+
CD	Cooling degrees	Days/month	+
BL	Average electricity price	$/kWh/month	−
TCL	Total connected load	kWh	+
AHDT	Designed heat loss	Btuh/ft^2/hr	+
ACDT	Designed heat gain	Btuh/ft^2/hr	+
DUM1	Apartments	0,1	?
DUM2	Churches	0,1	?
DUM3	Stores and Offices	0,1	?
DUM4	Light Manufacturing	0,1	?
DUM5	Motels	0,1	?

crimination between electricity used for space heating and the electricity used for space cooling in a particular month, the empirical analysis is restricted to months in which no electricity is likely to be used for space-cooling purposes [December (D), January (J) February (F), March (M)] or months in which little or no electricity is likely to be used for space heating purposes [June (JN), July (JY), August (AG), September (SP)].

Actual *heating degree-days* by month over a 12-month period at the exact location of the building are used as the measure of those climatic influences likely to affect the demand for space heating. This measure was presumably calculated by the Electrical Heating Association in accordance with standard practice. That is, the measure represents the difference between a base temperature of 65°F and the daily arithmetic mean temperature, given that the latter is less than the former.

Cooling degree-days are symmetrical in definition with heating degree-days. That is, they represent the difference between daily arithmetic mean temperature and a base temperature of 65°F, given that the latter is less than the former. However, for two reasons, the measure of this climatic variable actually employed in the empirical analysis probably does not indicate the actual climatic experience of each building as accurately as does the heating degree-day measure. First, the cooling degree-day measure is a simple average of experience over the four summers, 1969 to 1972. Thus, it is perhaps more an indicator of typical summer weather than of the actual weather during the months between 1963 and 1971, when observed electricity consumption in a

Table 2. Arithmetic Means and Standard Deviations of Original Values of All Variables Reported in Regressions ($n = 81$)

Variable	December	January	February	March	April	May
DD	714.9	1053.8	835.83	662.59	372.96	186.82
	(257.97)	(333.02)	(294.04)	(263.29)	(242.09)	(142.77)
CD	5.03	3.42	2.13	6.51	42.77	112.99
	(10.67)	(8.54)	(4.22)	(12.55)	(59.49)	(94.79)
BL	0.0144	0.0124	0.0130	0.0132	0.0141	0.0157
	(0.0105)	(0.0043)	(0.0048)	(0.0047)	(0.0056)	(0.007)
TCL	926.211					⟶
	(1359.76)					
AHDT	0.4639					⟶
	(0.2405)					
ACDT	1.7053					
	(0.8781)					
DUM1	0.0556					⟶
	(0.2303)					
DUM2	0.0778					⟶
	(0.2693)					
DUM3	0.5556					⟶
	(0.4997)					
DUM4	0.0333					⟶
	(0.1805)					
DUM5	0.1000					⟶
	(0.3017)					
kWh	134056	143550	141108	132427	110626	105373
	(237781)	(236390)	(222810)	(219056)	(180694)	(198233)
BL(1971)	.01517	.01455	0.1457	.01489	—	—

Variable	June	July	August	September	October	November
DD	44.19	6.06	3.8	37.14	179.03	466.29
	(61.87)	(13.66)	(9.60)	(49.08)	(132.75)	(213.66)
CD	271.67	377.58	348.01	223.3	61.86	8.96
	(138.31)	(138.16)	(140.10)	(132.55)	(75.55)	(14.38)
BL	0.0158	0.0159	0.0154	0.0151	0.0152	0.0150
	(0.0064)	(0.0073)	(0.0071)	(0.0064)	(0.0069)	(0.0071)
kWh	105820	108560	110480	108874	101939	111084
	(194500)	(204237)	(196588)	(187389)	(181965)	(187992)
BL(1971)	.01747	.01776	.01776	.01744	—	—

particular all-electric building occurred. Second, the locations at which the cooling degree-days were measured are not the exact locations of the buildings in the sample. Instead, the measurements were performed at the major city closest to each building. Since nearly all the buildings in the sample were in suburban or small town locations, it is likely, because of the heat islands generated by urban agglomerations, that the cooling degree-day measure is overstated. This would tend to bias downward estimates of the responsiveness of electricity consumption to changes in cooling degree-days.

The sole pecuniary variable for which results are reported is the *average price of electricity*. In order to define all prices in 1971 terms, this average price for all buildings in the sample except apartments was adjusted prior to empirical analysis by the commercial electricity price index of the U.S. Bureau of Labor Statistics.[18] The same procedure, using the residential electricity price index, was performed for the apartment buildings in the sample. This use of different price indices for apartments and all other buildings in the sample is somewhat arbitrary since some large apartment buildings might purchase electricity under commercial rather than residential tariffs.*

If one were interested in estimating the covariation of electricity consumption and electricity price, the use of the average price alone would usually be an inadequate representation of the price structure the decision maker faces. The reason is that electricity price is vector- rather than single-valued. Typically, electricity price schedules have fixed and variable components. From the perspective of the electricity consumer the unit price of electricity will decline with increasing consumption, both because of the presence of the fixed component and because nearly all schedules have variable components that decline with increasing consumption. The simple correlation coefficients for average unit prices between various months presented in Table 3 provide some insight for the data used in this study of the severity of the relation between price and quantity consumed. Given that price for any given quantity of electricity is unlikely to change very much from month to month or even over a year, the rather low simple correlation coefficients for pairs of months that adjoin each other can only be explained by differences in quantities consumed between the months.

Since it is not possible with existing techniques to include a vector as a term in a single estimating equation, the problem of representing electricity price reduces to finding a collection of terms that will provide a fair representation of the entire price schedule. As Taylor[19] observes, it is similar to the problem of finding a set of statistics that, for the purpose at hand, will adequately

* Except for cooling degree-days, all data are taken from the Electrical Heating Association case histories. Cooling degree-days are from U.S. Environmental Science Services Administration.[19]

Table 3.   Simple Correlation Matrix—Price Variables for Each Month

	BL(J)	BL(F)	BL(M)	BL(D)	BL(JN)	BL(JY)	BL(AG)	BL(SP)
BL(J)	1.000	.952	.927	.685	.790	.765	.797	.824
BL(F)		1.000	.950	.661	.776	.730	.791	.816
BL(M)			1.000	.644	.796	.738	.748	.800
BL(D)				1.000	.625	.606	.551	.579
BL(JN)					1.000	.940	.843	.909
BL(JY)						1.000	.841	.920
BL(AG)							1.000	.944
BL(SP)								1.000

represent a probability density function. In this chapter, it is assumed that average electricity price does provide an adequate representation. An analysis of the data indicated that average, and therefore marginal, prices were orthogonal to heating and cooling degree-days. Thus it seems likely that no bias was imposed upon the parameters of the climatic variables by failure to include more "moments" of the electricity price schedule. It is nevertheless true that the absolute magnitude of the estimated parameter for average price is biased upward since, if the consumption decision depends at least on marginal as well as average price, marginal price is less than average price. (See Crocker[5] for a description of a simultaneous equation model of residential demand for electricity.)

The current average price of a substitute energy source (natural gas) was introduced as an additional explanatory variable in the empirical tests. However, the estimated parameter for the variable was statistically insignificant, the presence of the variable did not seriously affect the standard error of any other variable, and its β-coefficient was very low. Therefore, results that include this variable are not reported. No other substitute energy source, for example, fuel oil, had its price used in the analysis.

Total connected load is a measure, in terms of the number of kilowatt-hours the stock can potentially draw, of the stock of electricity consuming capital goods. If each building maintains a constant interim temperature and if each item in the capital stock has a fixed coefficient for electricity use, this variable can be treated as a surrogate for output. That is, it represents the amount of heating or cooling required to achieve and maintain the interior temperature. The data available to this study on total connected loads represent the sum of the connected loads for specific purposes, for example, space heating and cooling, lighting, water heating, and ventilation. To the extent that uses of electricity other than for space heating and cooling do not vary systematically with climatic variations, the grouping of all electricity uses should be innocuous.

The designed heat loss and gain variables represent those structural attributes of the building, given the stock of electrical space-heating and -cooling equipment, that affect the quantity of electricity that must be consumed over a given time interval in order to maintain a desired indoor temperature. The values entered for each of these variables for each building were calculated in the following manner. First total design heat loss, HL, is

$$\text{HL} = (T_I - T_O) \sum_{i=1}^{4} A_i U_i$$

where T_I and T_O are respectively the desired indoor temperature and the outdoor temperature assumed by the builder. The outdoor temperature used by the builder for this calculation is generally 10 to 15°F greater than the lowest temperature in the recorded experience of a locale. The U's, or heat transmission coefficients, are the hourly rates of heat transfer through 1 sq ft of surface when a temperature difference of 1°F exists between the air temperatures on the two sides of the surface. Separate heat transmission coefficients were available for the walls, windows, floors, and ceilings of each building in the sample. The As are the number of square feet in each of the surface areas exposed to outside air or the ground.

Upon dividing equation (1) by $(T_I - T_O)$, one obtains a measure of the designed heat loss for the entire building that is independent of this outdoor air temperature. It is dependent only upon the size of the building and the insulating characteristics of the materials used in construction. A variable for designed heat gain can be formed by identical procedures:

$$\text{HLDT} = \sum_i A_i U_i; \qquad \text{HCDT} = \sum_i A_i U_i^*.$$

However, the Us in the heat gain calculations will differ slightly from those in the heat loss calculations because only the former account for relatively minor influences on heat transfer such as the number of persons expected to be in the structure.

Calculation and inspection of the values for HLDT and HCDT made it apparent they were highly collinear with the total connected load variable. Both were thus transformed in order to obtain the variable actually employed for estimation purposes.

$$\text{AHDT} = \frac{\text{HLDT}}{\sum_i A_i}; \qquad \text{ACDT} = \frac{\text{HCDT}}{\sum_i A_i}$$

The AHDT and ACDT thus represent the designed hourly heat loss and the designed hourly heat gain per square foot of surface area when a 1°F temperature difference exists between inside and outside air.

The remaining explanatory variables are dummies, the parameter values of which are defined relative to school buildings, the excluded building type. Without a detailed consideration of the uses to which the various building types are put, it was not possible to specify a priori the expected signs of the parameter values for each building type. Since a detailed analysis of this sort did not appear to contribute anything to the unique purposes of this study, the analysis was not undertaken.

Empirical Results

All ordinary-least-squares regressions reported in this section are multiplicative in accordance with the earlier specification of the Cobb-Douglas production function for transforming combinations of electricity and other inputs into a variety of outputs. Therefore, the coefficients of the explanatory variables are elasticities showing the percentage change in kilowatt-hours demanded with a 1% change in the explanatory variable.

Tables 4 and 5 present the major regression results. Although the year has been partitioned so that only the summer and winter months are given attention, it seems unlikely that neglect of the spring and fall months results in a significant loss of information about the effects of climatic change. During the spring and fall months, electricity used for controlling internal building temperatures will likely be used for both heating and cooling. In general, the heating and cooling degree-day variables were quite collinear, with simple correlation coefficients ranging from −.56 to −.84. When both variables were included in the same equation, the standard error of the previously included variable tended to increase dramatically. Any unidirectional change will increase one type of use and reduce the other. However, the results of the March, June, and September regressions indicate that there is substantially less responsiveness to variations in heating degree-days (March) or cooling degree-days (June and September) during these months than for other months for which regression results are reported. These relatively low elasticities are consistent with a relatively small absolute change in quantity demanded with respect to a one unit absolute change in degree-days. If it is true that there exists a range of temperatures (possibly 60 to 75°F) in which neither heating nor cooling activities are typically undertaken, regardless of electricity prices, then only temperatures lying outside this range would bring about an electricity consumption decision, even though any climatic deviation whatsoever from 65°F would be included in the degree-day calculation. Thus, it would appear

Table 4. Summary of Heating Regressions—Dependent variable: ln (KWF)

	January	February	March	December
ln(DD)	.262	.269	.098	.314
	(.107)**	(.113)**	(.094)	(.106)*
ln(BL)	−.250	−.714	−.978	−.522
	(.200)	(.211)*	(.230)*	(.149)*
ln(TCL)	.958	.950	.948	.971
	(.045)*	(.047)*	(.052)*	(.050)*
ln(AHDT)	−.250	−.247	−.261	−.385
	(0.84)*	(0.89)*	(.095)*	(.098)*
DUM1	−.604	−.596	−.571	−.534
	(.206)***	(.223)***	(.230)***	(.236)***
DUM2	−.494	−.676	−.529	−.644
	(.169)***	(.179)***	(.192)***	(.198)***
DUM3	.132	.089	.273	.182
	(.111)	(.117)	(.125)***	(.130)
DUM4	.266	.853	.456	.395
	(.219)	(.431)	(.248)	(.254)
DUM5	.190	.436	.303	.195
	(.152)	(.298)	(.173)	(.178)
constant	.417	−.247	−.035	.320
R^2	.938	.934	.926	.926
S	.337	.357	.383	.392
F	121.453	112.858	99.692	100.460
df	72	72	72	72

* The coefficient is significantly different from zero at the .01 level of the one-tailed t test.

** The coefficient is significantly different from zero at the .05 level of the one-tailed t test.

*** The coefficient is significantly different from zero at .05 level of the two-tailed t test.
(The numbers in parentheses are the standard errors of the estimated parameter values.)

in the regression results that, on the average over the observed temperature ranges for the months in question, electricity consumption for heating and cooling purposes was relatively unresponsive to changes in degree-days. This behavior seems particularly likely during the spring and fall months in the temperature climates where the buildings used in this study are located.

Except for the months of March, July, and August, the degree-day coefficients and their associated standard errors are remarkably similar to one

Table 5.   Summary of Cooling Regressions—Dependent variable: ln (KWH)

	June	July	August	September
ln(DC)	.248	.591	.672	.246
	(.115)**	(.192)*	(.197)*	(.104)**
ln(BL)	−1.331	−1.574	−1.243	−1.441
	(.271)*	(.298)*	(.311)*	(.336)*
ln(TCL)	(.951)	.905	.958	.924
	(.081)*	(.089)*	(.091)*	(.094)*
ln(ACDT)	.257	.337	.352	.282
	(.123)**	(.143)**	(.148)**	(.150)
DUM1	−.207	−.036	−.087	−.147
	(.325)	(.366)	(.383)	(.375)
DUM2	−1.141	−.928	−1.006	−.983
	(288)***	(.323)***	(.336)***	(.333)***
DUM3	.317	.479	.448	.403
	(.194)***	(.218)***	(.227)	(.224)
DUM4	.588	.853	.696	.771
	(.381)	(.431)	(.450)	(.442)
DUM5	.262	.436	.345	.184
	(.265)	(.298)	(.308)	(.304)
constant	−2.484	−5.547	−4.852	−2.800
R^2	.875	.846	.831	.837
S	.587	.662	.689	.680
F	55.820	43.893	39.381	40.675
df	72	72	72	72

* The coefficient is significantly different from zero at the .01 level of the one-tailed *t* test

** The coefficient is significantly different from zero at the .05 level of the one-tailed *t* test.

*** The coefficient is significantly different from zero at the .05 level of the two-tailed *t* test.

(The numbers in parentheses are the standard errors of the estimated parameter values.)

another. If substantial differences in typical temperature ranges exist over these months, the results would seem to indicate that, at least for this sample of buildings, there exists for temperatures below 75 or 80°F a uniformity of the responsiveness of electricity consumption to variations in climate. Since July and August are the warmest months of the year in most locales, the July and August coefficients indicate that this uniformity of response probably breaks down when the climate becomes uncomfortably hot. However, in no month does the proportionate change in electricity consumption exceed the proportionate change in degree-days, although, given the July and August results, it is quite possible that the proportionate change in electricity consumption comes closer to exceeding the same change for degree-days as the temperature increases after passing some unknown threshold. If this supposition is correct, it would imply that less electricity is consumed in cooler climates.

It should be noted that there does not appear to be any severe multicollinearity between the degree-day variables and the other explanatory variables. The simple correlation coefficients of Table 6 show that serious multicollinearity is quite unlikely. In addition, the deletion of one or more explanatory variables from the regressions did not have any significant effect on the standard errors of the degree-day variables; neither did the deletion of the degree-day variables have any significant effect on the standard error of any other explanatory variable.

Concluding Remarks

It has been earlier noted that, along with all the other qualifications embodied in this chapter, the magnitudes of the parameter estimates for degree-days are meaningful if, and only if, one is willing to assume that structural attributes and substitute energy sources are held constant. This cautionary note should now be extended to the uniformly positive signs of these parameter estimates. In general, a positive elasticity of derived demand with respect to climate will result in a positive climate elasticity of substitution of electricity for other inputs that help to maintain interior air temperatures. Thus, on balance, these other inputs are presumed to be complementary to electricity consumption as climate is altered. In a situation where oil, natural gas, and other fuels, as well as building materials, serve as obvious substitutes for electricity and

Table 6. Simple Correlation Matrix for January and July—Variables and All Other Independent Variables

	DD(J)	DD(JY)	CD(J)	CD(JY)	BL(J)	BL(JY)	TCL	AHDT	ACDT
DD(J)	1.000	.236	−.790	−.470	.063	.135	.002	−.067	.001
DD(JY)		1.000	−.279	−.562	−.058	.020	−.116	.027	.004
CD(J)			1.000	.534	−.071	−.111	.050	−.025	.007
CD(JY)				1.000	.108	−.028	−.004	−.026	−.021
BL(J)					1.000	.765	−.529	.179	.061
BL(JY)						1.000	−.485	.195	−.091
TCL							1.000	−.187	−.266
AHDT								1.000	.273
ACDT									1.000

where the per building stock of electricity-consuming capital equipment is relatively constant from one U.S. climatic region to another, this presumption seems somewhat questionable. In a model where simultaneous determination of energy forms and building materials is permitted, the signs attached to the parameters of the degree-day variables in an expression for the derived demand for electricity might well be negative.

Finally, it should be emphasized that use of a single climate elasticity may be seriously misleading. Choice behavior in colder climates with high heating degree-days and low cooling degree-days may differ from choice behavior in warmer climates with low heating degree-days and high cooling degree-days. As one moves from cold to colder climates, it may, for example, be that the elasticity of derived demand for electricity is negative implying that building materials and other means of space heating become progressively less costly relative to electricity. Conversely, as one moves from warm to warmer climates, cooling by electricity may become relatively less costly, implying a positive elasticity of derived demand with respect to climate. This possibility makes it worthwhile to partition by season and/or locale any data set used to implement a model of electricity demand. Otherwise, one obtains an elasticity estimate representing behavior in neither the cold nor the warm regions, but simply a weighted average of the two from which it is impossible to disentangle the separate contribution of each elasticity.

References

1. Anderson, K. P., "Residential Demand for Electricity: Econometric Estimates for California and the United States," *Journal of Business,* Vol. 46 (October 1973), pp. 526–553.

2. Anonymous, "Fuel Shortage Leads to Doubling of All-Electric Home Construction," *Washington Post,* (Sunday, July 28, 1974), pp. B-1, B-3.

3. Baxter, R. E. and Rees, R., "Analysis of the Industrial Demand for Electricity," *The Economic Journal,* Vol. 78 (June 1968), pp. 277–298.

4. Chapman, D., "Electricity in the United States," in Erickson, E. W. and Waverman, L., Eds., *The Energy Question: An International Failure of Policy,* Vol. 2, Toronto: University of Toronto Press, 1974, pp. 77–96.

5. Crocker, T. D., "A Simultaneous Equation Model of the Residential Demand for Electricity," in R. C. D'Arge, Ed., *Economic and Social Measures of Biologic and Climatic Change,* Washington, D.C.: U.S. Government Printing Office, forthcoming. In a simultaneous equation model of the residential demand for electricity, Crocker[5] includes the following expression to be estimated as part of the model.

$$P_E = \delta_O \chi_E \left(\frac{\Delta P_E}{\Delta \chi_E}\right)^{\delta_2} \xi$$

where P_E is electricity "price," χ_E is the quantity of electricity demanded, and ξ is an error term having the usual properties. The constant, δ_O measures the position of the price schedule in the positive orthant, and $\Delta P_E / \Delta \chi_E$, recognizing that most schedules are discontinuous, represents the average slope of the schedule in its intermediate ranges. The formulation recognizes that the price for any given quantity of electricity consumed may differ from one locale to another, that the consumer faces a price that varies with the quantity of electricity he consumes, and that the slope of the price schedule may change over space.

6. Fisher, F. M. and Kaysen, C., *A Study in Econometrics: Electricity Demand in the United States,* Amsterdam: North Holland, 1962.

7. Green, H. A. J., *Aggregation in Economic Analysis: An Introductory Survey,* Princeton: Princeton University Press, 1964.

8. Griffin, J. M., "The Effects of Higher Prices on Electricity Consumption," *The Bell Journal of Economics and Management Science,* Vol. 5 (Autumn 1974), pp. 515–539.

9. Halverson, R., *Demand for Electric Power in the United States,* Discussion Paper 73-13, Seattle: University of Washington Institute of Business Research, 1973.

10. Hicks, J. R., *The Theory of Wages,* New York: Peter Smith, 1948.

11. Houthakker, H. .S., "Some Calculations on Electricity Consumption in Great Britain," *Journal of the Royal Statistical Society, Series A (General),* Vol. 64, Part III (1951), pp. 359–371.

12. Houthakker, H. S. and Sheehan, D. P., "Dynamic Demand Analysis for Gasoline and Residential Electricity," *American Journal of Agricultural Economics,* Vol. 56 (May 1974), pp. 412–418.

13. Houthakker, H. S. and Taylor, L. D., *Consumer Demand in the United States,* Cambridge: Harvard University Press, 1970.

14. Johnson, S. R., et al., "Temperature Modification and Costs of Electric Power Generation,: *Journal of Applied Meteorology,* Vol. 8 (December 1969), pp. 919–926.

15. Landsberg, H., et al., *Resources in America's Future,* Baltimore: Johns Hopkins University Press, 1963.

16. Tansil, J., *Residential Consumption of Electricity: 1950–1970,* ORNL-NSF-EP51, Oak Ridge, Tenn: Oak Ridge National Laboratories, 1973.

17. Taylor, L. D., *The Demand for Electricity: A Survey,* unpublished manuscript, 1974.

18. U.S. Bureau of Labor Statistics, *Retail Prices and Indexes of Fuels and Utilities,* Washington, D.C.: U.S. Government Printing Office, 1971.

19. U.S. Environmental Science Services Administration, *Climatological Data: National Summary,* Washington, D.C.: U.S. Government Printing Office, 1971.

20. Wilson, J. W., "Residential Demand for Electricity," *The Quarterly Review of Economics and Business,* Vol. 11 (Spring 1971), pp. 7–22.

7

Climate and Household

Budget Expenditures

THOMAS D. CROCKER
Department of Economics
University of Wyoming
Laramie, Wyoming

There exists a long-standing and substantial anthropogeographical literature dealing with the influence of climate upon man and his activities. Much of this literature is descriptive in that it simply lists associations between climate types and the activities man undertakes. For an example of this literature see Maunder.[11] Those parts of the literature that do devote any attention to more abstract concerns generally concentrate upon perceived relations between climate and the broad structure of human culture and society.* A substantial emphasis is frequently placed upon the influence of climate upon man's preference ordering, both with respect to his preferred consumption patterns and his willingness to work. For example, the 18th century French jurist and philosopher, Montesquieu, asserted "The nations of hot countries are timorous like old men, the nations in colder regions are daring like youngsters."[12] In a more modern vein, the economic geographer Huntington has stated ". . . cultural habits which do not accord with the geographical environment either cannot establish a footing or else die out."[7] These broad-based perspectives on variations across climates in preference orderings and the want—satisfying capacity of various goods can be intriguing; however, whether strictly empirical or more speculative, they provide little, if anything, in the way of a systematic analysis permitting one to derive propositions susceptible to empirical test. Neither do they allow any reasonably precise indication of the economic values of alternative climates.

* For an interesting review of the intellectual history of this literature see Maršik.[10]

Except insofar as learning phenomena can be introduced, formal economic theory currently has very little to offer with respect to a systematic understanding of changes in preference orderings or tastes.* In general, value comparisons are considered to be definable only for exogenously given preference orderings. However, if one is willing to assume that the preference ordering of an individual is invariant over moderately short time intervals, economic theory does indeed provide an engine of analysis permitting one to establish empirically testable propositions about the behavior of man in different climates as well as the values he attaches to these climates. The fundamental purpose of the present paper is thus to illustrate, in an admittedly partial and incomplete fashion, how the economic theory of the consumer or household might be used to infer the values people attach to alternative climates.

Other than serving at the conceptual level as another good to which the consumer may attach value, climate does not play a role in the basic economic theory of consumer demand. Limited to a particular expenditure within a particular time interval and facing given prices, the consumer's problem in this basic theory is to select that combination of goods yielding him the highest utility. If his budget constraint does not permit satiation, he thus must maximize

$$U = U(z_1, z_2, \ldots, z_n)$$

subject to

$$\sum_{i=1}^{n} P_i z_i = y \tag{1}$$

where U is a strictly concave, twice-differentiable utility function, z_i is the quantity purchased of the ith good, and y is the consumer's money income. Necessary conditions for an optimum require that the budget constraint be satisfied and that the marginal rates of substitution for all pairs of commodities be equated to their prices.[6] Demand functions are readily derived from these so-called first-order conditions. Thus the optimal quantity of the ith good is a function of the price of the good, the prices of all other goods, and the consumer's income. The exact form of this function

$$z_i = z_i(P_1, P_2, \ldots, P_n, y) \tag{2}$$

depends on the consumer's preferences. By totally differentiating the first-order

* See Fisher and Shell,[3] pp. 1–26, for one treatment of how taste changes might be treated in consumer theory and welfare propositions established that systematically incorporate them. Fisher and Shell base their welfare criteria upon the comparison of the present and past constraints the consumer faces rather than his present and past utilities. Other formal economic treatments of taste changes are available in Rothenberg[18] and von Weizsacker.[19] Gintis,[4] pp. 273–277, presents an interesting discussion from the perspective of a radical economist of the analytical implications of the assumption of exogenously given consumer preference orderings.

conditions, one obtains expressions for the rate of change of the quantity of z_i demanded with respect to changes in income, in the price of the good, or in the price of any other good. Those expressions that relate to price changes can be decomposed into substitution effects and income effects

$$\frac{\partial z_i}{\partial P_j} = \left(\frac{\partial z_i}{\partial P_j}\right)_{U_{\text{constant}}} - z_j \left(\frac{\partial z_i}{\partial y}\right) \qquad (i, j = 1, 2, \ldots, n) \qquad (3)$$

In the absence of inferior goods, one can show from this formulation that for a single good the substitution effect is negative, that the demand function is homogeneous of degree zero in all prices and income, and that quantity demanded and price vary inversely.

For our purposes, the most noteworthy feature of this basic approach to consumer theory is that complete specification of the demand function for a single good requires the inclusion of the consumer's money income and the prices of *all* goods. If climate is a good in and of itself and/or if it influences the utility to be obtained from other goods, then failure to include it or its price in any attempt to estimate the demand function for another good can bias the estimated results.* That is, the estimated impact of income or the prices of goods that are included upon the quantity demanded of the good for which the demand function is being estimated could be exaggerated or deflated. To provide some motivation for others to reflect upon the ways in which climate might effectively be introduced into empirical demand studies, it therefore seems worthwhile to discuss climate as a good valued in and of itself and as a possible influence on the demand for other goods.

With the exception of a fair amount of work on the household demand for energy, climate is nearly always given short shrift in the empirical demand literature of economics. Apart from the energy studies referenced in Chapter 6 by author in this volume, a literature search turned up only two empirical household demand or expenditure studies in which measures of climate were explicitly introduced.[2, 5] However, studies of the demand for individual goods and studies of demand systems, in which an attempt is made to account for the prices of all goods, frequently find significant interregional differences in expenditure patterns.† Attempts to explain these differences do not consider the possibility that climate may contribute to an explanation.

The objectives of this chapter are threefold. First, I endeavor to outline some of the theoretical effects of variations in climate on consumer behavior when differences in climate do no more than influence the cost of using goods capable

---

* When actually estimating demand functions or expenditure functions, for individual goods, the necessity of including the prices of all goods is avoided by imposing *a priori* separability conditions that make the prices of various collections of goods independent of other collections. See Muth[15] for a discussion of these conditions.

† For a thorough review of the theory and estimation of demand systems, see Powell.[16]

of providing utility for the consumer. Thus, if growing roses is an activity or good from which a consumer derives utility, he will have to expend more valuable resources undertaking the activity in Laramie, Wyoming, than in Riverside, California. Second, this outline is expanded to permit the consumer to obtain utility directly from climate while still having climate affect the cost of other activities from which he obtains utility. The chapter closes with a brief discussion of the possible empirical implications of the theoretical formulations discussed earlier. Those readers familiar with the theoretical literature in the economics of residential location will readily recognize that the entire paper draws heavily for its basic contributions upon this literature.[1, 14]

In what follows climate is viewed as a combination of long-term weather attributes such as ambient temperature, humidity, cloudiness, and their frequency over time. Indeed, since climate is the major determinant of the composition of flora and fauna in a locale and is an important determinant of land forms, climate might well be viewed as the entire natural environment of a locale. In spite of this, I assume throughout that each consumer or household views some combination of the aforementioned climatic attributes as the "ideal" climate. Any deviations from this unconstrained ideal are viewed as an increase in climate severity by the consumer. The consumer prefers less severe to more severe climates. Moreover, the climate and all goods, unless otherwise distinguished, are each assumed to fulfill the conditions for a composite good. That is, it is assumed that suitable indices can be constructed such that climate or collections of goods can be described by a scaler rather than a vector.

Climate as a Determinant of the Cost of an Activity

Let $U = U(z, x)$ be a strictly concave, twice-differentiable utility function increasing in z and x. Both z and x are quantities over which the household is able to exercise discretionary control. The quantity z represents consumer expenditures on goods or activities the unit cost of which is not sensitive to variations in climate. The quantity x is thus a composite of those activities having unit costs sensitive to variations in climate.

This utility function is maximized subject to a budget constraint of the form:

$$y = z + K(c)x + V(c) \tag{4}$$

Here y is defined as before, z is now total consumer expenditures on climate-insensitive activities; K a function of climate, c is the unit cost of x; and V is any rental costs due to climate alone, that is, the capitalized value of climate in any immobile and durable assets such as land the consumer might use. Implicit in $K(c)$ is the household production function approach to the problem of the consumer.[13] That is, the consumer is perceived as purchasing various goods on the open market that he combines to produce the activities that enter his utility function. The cost of producing any given magnitude of a particular activity may vary with climate. Both $K(c)$ and $V(c)$ are given exogenously to the consumer.

Utility maximization yields the constraint condition equation (4) and the following additional first-order conditions:

$$\frac{\partial U}{\partial z} = \lambda = 0; \qquad \frac{\partial U}{\partial x} = \lambda K\,(c) = 0$$

$$\frac{\partial V(c)}{\partial c} = -x\,\frac{\partial K(c)}{\partial c} = 0 \tag{5}$$

The first two expressions in equation (5) imply:

$$\frac{\partial U}{\partial z} = \frac{\partial U/\partial x}{K(c)} \qquad \text{or} \qquad \frac{\partial U/\partial x}{\partial U/\partial z} = K(c) \tag{6}$$

which states that, apart from climate, the consumer undertakes those amounts of climate-insensitive expenditures, z, and climate-sensitive activities, x, such that the marginal utility per dollar spent is the same for all activities. Note that the shadow price, λ, of the climate-sensitive activities is not a constant but is determined by the effect of climate upon the cost of undertaking these activities. If some combined index of the prices embodied in z and the costs of x is influenced by climate, it is equivalent to a change in the consumer's cost of living. That is, the utility the consumer can obtain for a given money income is affected. To repeat, in this formulation, climate is not a direct object of preference but rather a determinant of the magnitude of a utility-generating activity to be obtained from the use of a particular combination of market-purchased goods. In Figure 1 this phenomenon is illustrated by a shift from MC_0 to MC^1, of the marginal cost of producing the utility generating activity.

The $\partial U/\partial x$ term in equation (6) can be interpreted as the marginal utility the consumer obtains from an additional unit of the climate-sensitive activity. In Figure 1 the marginal utility function is represented by MU. Assume, for concreteness, that x in Figure 1 measures the interior temperature of a building. The shift of the marginal cost function from MC_0 to MC^1 is intended to represent the effects of a change in outdoor temperature that makes it more costly to maintain an interior temperature differing from the outdoor temperature. The positioning of the intercepts of the cost curves is arbitrary, although it is not implausible that in the case of interior cooling some inputs must be applied to keep interior temperature down to the level of the outdoor temperature. For interior space heating, the intercepts of the cost curves might go through the horizontal axis, since the interior temperature of most buildings is likely to be warmer than the outdoor temperature, even in the absence of any space-heating efforts.

Given the cost functions for maintaining various interior temperatures under

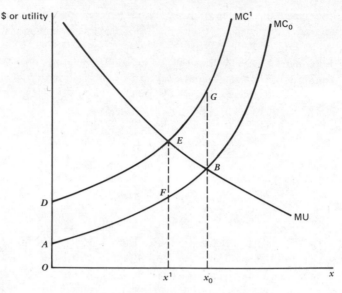

Figure 1. The influence of climate.

the new and old climates, the consumer of Figure 1 chooses an interior temperature of x_0 under the old climate and x^1 under the new climate. Under the old climate, the cost of maintaining the utility-maximizing interior temperature is the sum of the incremental costs of attaining x_0. This is given by the area $OABx_0$. The corresponding area with the new climate is $ODEx^1$. The difference between these two areas is the change in the individual's interior space-heating expenditures because of the change in climate. Since $OAFx^1$ is common to the two areas, the change in expenditure is given by the difference between $ADEF$ and x^1FBx_0. $ADEF$ is simply the change in cost necessary to maintain the interior temperature at x^1, and $ADGB$ is the change in cost necessary to maintain the former utility-maximizing temperature of x_0. It should be noted that the only circumstances in which $ADGB$ would be a correct measure of the value of the change to the consumer is when the MU function could be disregarded. That is, if and only if the level of the interior temperature were completely unresponsive to changes in the cost of providing it, so that the MU function exactly coincided with the GBx_0 line, would a measure of this sort actually measure value to the consumer. Otherwise it neglects the change in value to the consumer expressed in the movement along the MU function as the marginal cost functions shift in response to the climate change. If the interior temperature level demanded is not completely unresponsive to variations in the cost of providing it, the measure will fail to capture the value of the sum of the excess of the utility of the degrees of temperature no longer maintained over the sum of the costs of producing these degrees.

Simple comparisons of the aforementioned areas of Figure 1 make it clear that it is quite possible to observe zero or quite small changes in expenditures and still have the consumer experience substantial changes in utility. All that is required is the equivalence or near equivalence of $ADEF$ and x^1FBx_0. By its very nature, an expenditure measure is a measure of unit costs multiplied by the number of units. With a change in climate and its consequent effects upon the costs of providing interior temperatures, the consumer may adjust the price he is willing to pay for an additional hour of a given interior temperature or an additional degree of interior temperature, as well as the actual interior temperature he desires. If the change in climate is unfavorable, the utility-maximizing interior temperature will deviate farther from the consumer's unconstrained ideal, and the value the consumer attaches to a given movement toward the ideal will increase. According to whether the proportionate movement from or toward the ideal is greater or less than the proportionate change in willingness to pay, observed expenditures on interior temperature maintenance may increase or decrease, respectively. An increase suggests that the demand elasticity for the activity is less than unity, while a decrease suggests a demand elasticity greater than unity.

If the formulation of the consumer's problem being pursued here is at all reasonable, the ambiguities and complexities raised by introducing climate into studies of the demand for market-purchased goods perhaps explain why climate has been almost entirely disregarded in such studies. Thus, for example, the studies that deal with the demand for fuels are investigating the effect upon the consumer's willingness to pay for fuels of changes in utility-maximizing levels of interior temperatures, quantities of food processed, or a variety of other goods or activities directly entering the consumer's utility function.* The Marshallian rules of derived demand apply.[9]

Thus the extent to which the demand for a good such as fuel will be responsive to a change in climate will vary directly with the price elasticity of demand for the activity the fuel permits, the price elasticity of supply for substitute goods, the relative ease of substitution of other goods for the fuel in question, and the relative importance of the fuel in providing the valued activity. Therefore, a change in climate that increases the ease of maintaining a desired interior building temperature with a given type of fuel will, in order to minimize the costs of providing the desired temperature, cause the relative use of the

---

* By referring to willingness to pay, I am assuming that if an individual is to prevent a change in climate, he must pay to do so. His willingness to pay expresses the maximum sum he would pay if costless processes existed by which he could convey this willingness to those causing the change in climate.

fuel to increase. Since the relative cost of the desired interior temperature has decreased, this temperature will more frequently be maintained. However, these effects do not necessarily imply that the absolute quantity demanded of the fuel in question will increase. For such an increase to occur the increased quantity demanded, generated by the combined effects of the substitution of the fuel for other goods and the expansion of the activity the fuel provides, must outweigh the saving induced by the enhanced contribution of a given fuel quantity to the maintenance of interior temperature. That is, the utility gained by the improved temperature situation and the reduced expenditures on other goods must outweigh the utility gained by the improved maintenance capacity of any given quantity of the fuel. If the market price of the fuels is unchanged, it is easy to see that a given climate change can increase or decrease expenditures on fuels, according to whether the change increases or decreases the quantity of fuels demanded.

The discussion of Figure 1 and of the first two first-order conditions of equation (5) entirely neglects the third expression in equation (5). This expression means that in the equilibrium climate the consumer is unable to loosen his budget constraint by any change in the climate to which he is currently subject. The term $-x[\partial K(c)/\partial c]$ is the change in expenditure on climate-sensitive activities resulting from any increase in climate severity, while the $\partial V(c)/\partial c$ term is the change in capitalized rents due to a change in climate. In general, any increase in climate severity results in a decline in climate rents. Since $\partial V(c)/\partial c < 0$, expenditures on climate-sensitive goods must then increase with increasing climate severity if the third condition in equation (5) is to be satisfied. This is fairly obvious since if all goods were climate-insensitive, but climate rents declined with increasing climate severity, any increase in climate severity would increase the consumer's real income. Everybody would then try to live in as severe a climate as possible. The third condition in equation (5) ensures that with increasing climate severity the consumer trades off the declining climate rents he must pay against increasing expenditures on climate-sensitive goods. This third condition in equation (5) specifies the climate in which the consumer maximizes his utility. Since this equilibrium choice of climate is determined in part by the relative costs of various activities, and since the equilibrium mix of activities chosen by the consumer influences the quantity of market-purchased goods the consumer uses, the consumer's willingness to pay for market-purchased goods cannot be determined independently of climate.*

The third condition in equation (5) may be used to analyze the effects of an exogenous change in any of the factors of equation (4). For example, the influence of money income on choice behavior is frequently of interest. Upon

* The difficulties this raises for estimating income-compensated demand functions are very ably treated in Rosen.[17]

totally differentiating the third condition of equation (5) by income and rear-ranging, one obtains

$$\frac{dx}{dy}\left(\frac{1}{x^2}\right)\left(\frac{\partial V(c)}{\partial c}\right) - \frac{1}{x}\left(\frac{\partial^2 V(c)}{\partial c^2}\right) = \left[\frac{\partial^2 K(c)}{\partial c^2} + \frac{1}{x}\left(\frac{\partial^2 V(c)}{\partial c^2}\right)\right]\frac{dc}{dy} \qquad (7)$$

As Muth shows,[14] in order for equation (5) to be a maximum, it must be that

$$\frac{\partial^2 K(c)}{\partial c^2} + 1\frac{\partial^2 V(c)}{\partial c^2} \geq \frac{dx}{dc}\left(\frac{1}{x^2}\right)\left(-\frac{\partial V(c)}{\partial c}\right) > 0 \qquad (8)$$

By assumption, $dx/dc > 0$. If $dx/dy > 0$, so that climate-sensitive goods are normal goods the quantity demanded of which increases with increases in income, then the term in brackets on the right-hand side of equation (7) will be positive. The sign of dc/dy, the change in the severity of the climate to which the consumer subjects himself with respect to variations in income, will thus depend on only the left-hand side of equation (7). If dx/dy is fairly large so that increases in the consumer's income result in substantial increases in the quantity of climate-sensitive goods he consumes, and if marginal rents decline slowly with increases in climate severity so that $\partial^2 V(c)/\partial c^2$ is small, then the left-hand side of equation (7) will be positive. This then means that dc/dy will be positive, implying that climate severity and income vary directly. Conversely, if dx/dy is small and $\partial^2 V(c)/\partial c^2$ large, then dc/dy will be negative, implying that increases in income cause the consumer to search out a less severe climate. Since the unit cost of an x activity is a function of climate, any income-induced changes in the climate the consumer chooses can alter the price of the activity and thereby influence the consumer's demand for the market-purchased goods that permit the activity. In essence, there exists a price effect of a change in income.

Climate as an Object of Preference and as a Determinant of Activity Cost

In this section, in addition to continuing to recognize that climate may influence the costs of undertaking various activities from which utility may be obtained, I introduce climate as a direct object of preference. That is, climate is permitted to enter the utility function directly. This amounts to a recognition that some individuals may find a particular climate attractive for its own sake. For example, a sunny, dry, and cool climate may be considered invigorating but an overcast, damp, and cool climate may seem depressing.

Let the consumer's strictly concave twice-differentiable utility function be $U = U(z, x, c)$, where the arguments are defined as in the previous section. The function is presumed to be increasing in z and x and decreasing in c. The consumer's problem is to maximize this function subject to the budget constraint in equation (4). Necessary first-order conditions for utility maximization still include equation (4) and the first two expressions of equation (5). All the remarks made about these first two expressions in the previous section are therefore applicable to this section. However, the third expression, which specifies the equilibrium climate for the consumer, is altered to become

$$\frac{\partial U}{\partial c} = \lambda \left[\frac{\partial K(c)x}{\partial c} + \frac{\partial V(c)}{\partial c} \right] = 0 \qquad (9)$$

Dividing equation (9) by the first expression of equation (5), one obtains

$$\frac{\partial U/\partial c}{\partial U/\partial z} = \frac{\partial K(c)x}{\partial c} + \frac{\partial V(c)}{\partial c} = 0 \qquad (10)$$

By assumption $\partial V/\partial c$ is negative and $\partial V/\partial z$ is positive, thus implying that the right-hand side of equation (10) must be negative. Therefore, since $\partial V(c)/\partial c$ is also negative, the absolute value of the second term on the right-hand side must exceed the first. If the schedule of climate rents is given exogenously to the consumer, this then implies that those consumers for whom the cost of climate-sensitive activities increase least rapidly with increases in climate severity will choose climates more severe than other consumers. In short, those consumers who, at the margin, have the least to lose by living in the most severe climates will do so.

As in the previous section, the influence of income on the equilibrium degree of climate severity the consumer will choose depends on the effect of changed income on the demand for climate-sensitive activities relative to the behavior of climate rents with respect to changes in climate. That is, the consumer's response will again depend on the tradeoff between expenditures on climate-sensitive activities and expenditures on climate rents.

When climate is both an object of preference and an influence upon the cost of obtaining utility-generating activities, additional insight about the influence of climate upon consumer expenditures can be had by employing the indirect utility function.* To establish the indirect utility function, one simply maximizes the direct utility function, $U(.)$, subject to the budget constraint, y, and then substitutes the demand functions corresponding to the arguments of the utility function for these arguments. Thus, for our purposes, consumer utility in a given climate can be expressed as

$$W(c) = W[K(c), c, y - V(c)] \qquad (11)$$

where W is the consumer's utility in his utility-maximizing climate, and $y - V(c)$ is his money income net of his rental costs in his utility-maximizing cli-

_____

* The technical properties of the indirect utility function are set forth in Lau.[8]

mate. For obvious reasons, W is increasing in $y - V(c)$ and decreasing in $K(c)$ and c.

If the consumer is able to choose the degree of climate severity to which he subjects himself, his equilibrium set of climate-sensitive activities will be such that he is unable to increase his utility by choosing any other degree of climate severity. This implies that at the equilibrium degree of climate severity there exists a relation between $K(c)$, the unit cost of climate-sensitive activities, and $W^*(c)$, the consumer's level of utility at the equilibrium degree of climate severity, as well as climate severity itself and the consumer's residual income net of climate rents.

$$K(c) = f[W^*(c), c, y - V(c)] \qquad (12)$$

Clearly, if one is interested in $K(c)$ and its effect on the demand for market purchased goods, one must establish exactly how $W^*(c)$ is determined. If $W^*(c)$ is determined only by the individual consumer's equilibrium degree of climate severity and other variables, no special problem is presented. One requires information only on this equilibrium degree of severity and the values of the only variables at this equilibrium degree in order to determine $K(c)$. However, the information required is substantially greater if there occurs a change in c, for example, that affects the utility of most consumers. From equation (11), a change in climate severity implies a change in each consumer's utility level and thus a change in each consumer's equilibrium level of utility. From equation (12) a change in the equilibrium level of utility implies a change in the consumer's cost of obtaining climate-sensitive activities and thereby a change in the consumer's willingness to pay for market-purchased goods that contribute to these activities. In effect, with a change in climate that affects the utility of a number of consumers, the cost of a climate-sensitive activity to the consumer continues to depend on the degree of climate severity to which he subjects himself. However, in addition, it now depends on the changes in climate severity elsewhere and the adjustments other consumers make in response to these changes. If, and only if, it is possible for consumers to make adjustments to climate changes such that the equilibrium level of utility is unaffected for each and every consumer can the influence of other consumers upon each consumer's $W^*(c)$ be disregarded in the determination of $K(c)$.

Summary and Conclusions

This paper has attempted to characterize and describe some of the implications of two ways of introducing climate into formulations of the consumer's problem. When the consumer employs market-purchased goods to obtain activities from which he derives utility, failure to consider the impact of climate upon the cost of obtaining these activities could cause serious specification errors in studies that purport to estimate the demand for market-purchased goods. These specification errors might be even more serious if climate is both an object of preference as well as a determinant of the relative costs of various activities. In this case, unless each and every consumer can make a set of adjustments to a change in climate that leaves his equilibrium level of utility unaffected, each consumer's cost of obtaining a particular activity, and thereby his willingness to pay for market-purchased goods that contribute to the activity, will be influenced by the choice behavior of other consumers. Thus, in order to determine the demand for a single market-purchased good, one may have to solve a general equilibrium problem if climate plays any part in consumer behavior.

References

1. Alonso, W., *Location and Land Use,* Cambridge: Harvard University Press, 1964.
2. Attaran, K., *Inter-Metropolitan Area Budget Cost Determinants: An Econometric Analysis,* Ph.D. dissertation submitted to the Department of Economics, University of Southern California, August 1973.
3. Fisher, F. M. and Shell, K., *The Economic Theory of Price Indices,* New York: Academic Press, 1972, 1–26.
4. Gintis, H., "Consumer Behavior and the Concept of Sovereignty: Explanations of Social Decay," *The American Economic Review,* Vol. 62 (May 1972), pp. 267–278.
5. Haworth, C. T. and Rasmussen, D. W., "Determinants of Metropolitan Cost of Living Variations," *Southern Economic Journal,* Vol. 40 (October 1973), pp. 183–192.
6. Hicks, J. R., *Value and Capital,* Oxford: Clarendon Press, 1946.
7. Huntington, C., *Mainsprings of Civilization,* New York, 1945, p. 610.
8. Lau, L. J., "Duality and the Structure of Utility Functions," *Journal of Economic Theory,* Vol. 1 (December 1969) pp. 374–396.
9. Marshall, A., *Principles of Economics,* 8th ed., London: MacMillan and Co., 1961.
10. Maršik, M., *Natural Environment and Society in the Theory of Geographical Determinism,* Acta Universitatis Carolinae Philsophica et Historica Monographia XXXI, Prague: Charles University, 1969.
11. Maunder, W. J., "National Econoclimatic Models," in J. A. Taylor, Ed., *Climatic Resources and Economic Activity,* New York: John Wiley, 1974.
12. Montesquieu, C., *The Spirit of Laws,* New York: Harper and Row, 1947, p. 245.
13. Muellbauer, J., "Household Production Theory, Quality, and the 'Hedonic Technique'" *The American Economic Review,* Vol. 64 (December 1974), pp. 977–994.
14. Muth, R. F., *Cities and Housing,* Chicago: University of Chicago Press, 1969, p. 25.

15. Muth, R. F., "Household Production and Consumer Demand Functions," *Econometrica*, Vol. 34 (July 1966) pp. 699–708.

16. Powell, A. A., *Empirical Analytics of Demand Systems*, New York: Lexington Books, 1974.

17. Rosen, S., "Hedonic Prices and Implicit Markets," *Journal of Political Economy*, Vol. 82 (January/February 1974) pp. 34–55.

18. Rothenberg, J., "Welfare Comparisons and Changes in Tastes," *The American Economic Review*, Vol. 43 (December 1953), pp. 885–890.

19. C. C., von Weizsacker, "Notes on Endogenous Changes in Tastes," A paper delivered to the Second World Congress of the Econometric Society, London, 1970.

8

Climate, Wages, and Urban Scale

IRVING HOCH
Resources for the Future
Washington, D. C.

Wage rates for the same work vary by locale. Possible explanations include disequilibria between locales, usually presumed to be short-run in nature, and long-run differences in the "quality of life" between locales which lead to compensatory differences in wage rates. ("Quality of life" is broadly defined here to include both pecuniary and nonpecuniary, or "psychic," components.) In the case of disequilibrium, workers can be expected to move from lower to higher paying areas, which will raise wages in the first set of places and lower them in the second, until equilibrium ultimately is established. But equilibrium does not necessarily involve equality of wage rates, because additional compensation is needed to retain employees in places with a higher cost of living, as measured by an index based on a market basket of traded goods, or in places with less pleasant environments, whose quality characteristics are not included in the conventional cost-of-living index.

It is plausible that climate is a factor in quality of life, and hence, affects wage rates. Places with good climate should have lower wages than places with bad climate. The higher wages in the latter places compensate both for increased money expenditures for additional heating or cooling expenses, affecting not only housing, but clothing, transportation, and perhaps medical expenditures, as well, and for increased "psychic" or nonmarket costs, expressing workers' preferences about the weather. Everybody not only talks about the weather, but does something about it, by taking it into account when making location choices. Hence, if climate varies, other things equal, there should be corresponding changes in real income that can be inferred from the behavior of money income in the form of wages.

This chapter applies and tests these notions by relating wages to climate variables and nonclimate sources of wage differentials employing a basic equation of the form:

$$W = a + b\mathbf{X} + c\mathbf{Z} \qquad (1)$$

where $\mathbf{X}$ is a vector of climate variables and $\mathbf{Z}$ of nonclimate variables, W denotes money wage, and a, b, and c are parameters to be estimated. In practice, W is a vector of money wages obtained in specific occupations for which wage data have been secured, yielding a set of equations of the same form, with vectors of parameters to be estimated. Climate variables employed in equation (1) were summer temperature, winter temperature, precipitation, the squares of the temperature variables, and the interaction of summer temperature and precipitation. Wind velocity was also employed in much of the work. Nonclimate variables were specified to include measures of urban scale (both population size and density), racial composition, population growth, and regional dummy variables, there being reason to expect that all of these variables affect wage rates. Thus, available evidence indicates that the cost-of-living increases, and that nonmarket components of the quality of life decline, on net, with urban scale in such items as rent and transportation expenditures, under the first heading, and in disutility from pollution and congestion, under the second. Such cost increases appear to outweigh any benefits of increased scale.*

Regional dummy variables can account for disequilibrium conditions, and for differences in cost of living by region; in particular, there is evidence of a pronounced and persistent cost-of-living differential between North and South. Some of that differential may be related to racial composition, and the possibility of either discrimination or disequilibrium as a function of race led to the inclusion of percent black as an explanatory variable. Finally, at a later stage in the investigation, population growth was introduced into some of the equations employed, under the hypothesis that rapid growth would be associated with above-average wages. The variable did not have much explanatory power, so it was decided not to introduce it in the earlier work, under the rationale that the costs would exceed the expected limited benefits.

The primary goal of this investigation was to estimate the effect of climate change on real income. That climate enters utility functions is, of course, a commonly entertained hypothesis. Thus, Eugene Smolensky[13] has recently argued that workers will accept a lower money wage to be in places with pleasant weather, and since warm weather is preferred to cold, much of the shift of industry and population to the south and west can be explained by cli-

* Some empirical estimates of net costs with scale appear in Hoch[6]; those results are summarized in Ref. 12.

mate preferences. Oded Izraeli[7] has estimated an equation similar to equation (1) by employing a number of quality of life indicators, including January temperature as his measure of climate. (His sample coverage was also limited to one occupational group—laborers.) The content of this chapter was foreshadowed in an earlier study when I noted[4] "there may be equilibrium differences in wages because natural conditions affecting consumer utility will vary between locale, e.g. 'good' versus 'bad' weather, differences in risk of earthquake, flood or hurricane, etc."

To attain the goal of this investigation, each of the set of equations fitting the form of equation (1) was estimated by single equation regression, and an aggregate equation was derived. Then, given specified changes in the **Z**, wages for *Y* could be determined before and after the climate change, and the ratio of wages "after" to wages "before" could be taken as an index of the cost or benefit of the climate change. The goal here is subsumable under the even more basic goal of establishing the existence of a significant relationship between climate and wages, reflecting quality-of-life effects, and thus developing substantive evidence on the interrelation between man's economic activities and the natural environment. Under this latter, more general heading, there were some side investigations of reverse causation involving the impact of urban scale on climate, interpreted as reflecting the urban heat island and the impact of particulate air pollution. This work was frankly experimental, and care must be exercised in interpreting results. Yet the results seem good enough to merit further research along these lines.

The Economic Model

A fairly elaborate model can be presented as rationale and underpinning for equation (1). We can begin the discussion by considering whether some or all of the benefits of good climate are capitalized into land values rather than appearing as an offset to wages, so that equation (1) would then understate the value of climate differences.

Certainly, it is well established in economic theory that profits often are capitalized into land value, and it is plausible that profits can be generated by good climate. But it is also well established in the theory of wages that differences in working conditions—hazard, or unpleasant conditions, including climatic differences—will generate a supply response affecting relative wages. Under fairly plausible assumptions, it seems possible to have both conditions hold at the same time.

Assume that there are two regions, and that region A has better climate than region B, in terms of consumer preferences. As an initial approximation, we can posit two perfectly elastic labor supply curves for the respective regions, with the A curve below the B, the difference between the curves representing the value of the climate differential. If aggregate demand for labor happened to be the same in both A and B, at equilibrium the population in A would be greater than that in B. An implicit key assumption is that the supply of land is perfectly elastic. This case might be taken as representing a world without cities, where wild land can be converted to farm land at constant cost. Hence, the relatively greater population in region A does not increase land values, which in turn would cause equilibrium wages in A to rise to cover higher rents, and reduce the wage differential between regions at equilibrium.

Next, assume a world with cities. Say a limited number of geographic locales have comparative advantage as city sites. If economic activity is focused on the urban center, with land at the center obviously in fixed supply, and if the city generates profits, such profits will be capitalized into urban land values. The profits will decline with distance from the center until at the city limits the cost of converting farm land to urban land just equals the returns from that conversion. As cities grow in population size, greater demand for land will increase rents and land values, and correspondingly, there will have to be an increase in equilibrium wages. Workers must be paid more to compensate for higher rents and, by multiplier effects, for a higher cost of living generally. In addition, pollution and congestion costs tend to increase with city size.[5, 7] Given the previously noted evidence that these costs outweigh any benefits of scale, the latter reflecting increased specialization and division of labor, we can generalize our initial labor supply case to one involving positive slope in the longest run. Thus, any urban area expanding or contracting its total employment will be faced with corresponding increases or decreases in the money wage rate, and all urban areas lie on a conventionally sloping labor supply curve. This generalization of the initial labor market situation is diagrammed as Figure 1, using the argument that workers must receive higher wages to cover higher cost of living (including lower environmental quality) as city population increases, and under the simplifying assumption that city population size is a constant multiple of employment. In Figure 1, S_A and S_B are the supply curves that hold for any cities in the respective regions. (We can think of the wage for labor quantity near zero as the farm wage.)

There are a large number of cities in each region, and we treat each city as analogous to a firm and develop an analysis of labor market equilibrium corresponding to firm production theory. Each city has its own production function for aggregate output, and as a consequence, a demand for labor equal to labor's value of marginal product (VMP). For each city, the intersection of VMP and the supply of labor determines the equilibrium wage and quantity of labor (and population as a scalar of labor), accounting for the distribution of city size by region.

The equilibrium points observed by the statistical investigator lie on the respective supply curves, and the introduction of climate variables accounts for

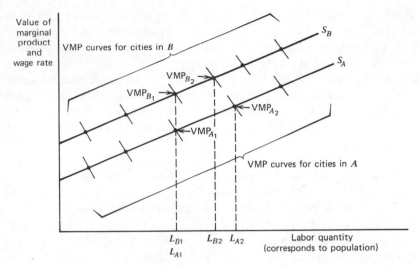

Figure 1. Labor market equilibria for cities in two regions with climate differences.

the shift that occurs between S_A and S_B. In Figure 1, two specific cases in each region are labeled for illustrative purposes. City 1 in B has the same population size as City 1 in A (L_{B1} and L_{A1}) by virtue of a higher VMP curve as well as a higher supply curve. The difference in wages between the two cities of course equals the climate benefit to Region A. City 2 in B has essentially the same VMP curve as City 2 in A, but the supply curve differences yield both a lower population and a higher money wage for the former city relative to the latter. Because cities in A are larger for the same VMP than those in B, those cities in effect will have obtained higher land values by virtue of their larger size. But if we assume that land can be converted from wild land to farm land, and from farm land to urban land, at fixed costs, then the S_A curve will not shift upward as a consequence of higher rents, that is, the benefits of better climate are not directly capitalized into land values, which in turn would reduce the wage differential. Hopefully, the real world will be close enough to this case of perfectly elastic land supply at the urban periphery to rationalize the wage estimation approach used here. (On an empirical level, it seems worth noting that land in urban use comprises about 1% of total land in the United States.)

In the statistical application of the model, with observations on a set of urban areas, we will have a corresponding number of VMP curves whose intersections with the supply curves will have been the source of the observed points. Instead of two supply curves, there will be a family of curves whose shifts are accounted for by the climate variables. In addition, there may be shifts accounted for by regional effects, perhaps explainable by labor market disequilibrium at a regional level. We can think of the population in urban areas at a given point in time as having been determined by the process of Figure 1 over a long period prior to that time, so that resultant populations are essentially exogenous. However, if all the VMP curves in a given region are above equilibrium levels, wages in turn will lie above the long-term equilibrium, yielding a positive coefficient for the regional dummy variable. Regional dummies should be useful not only in accounting for such disequilibrium effects, but also in avoiding attribution to climate variables of effects that are really attributable to region.

Single equation regression can be defended because population is essentially exogenous. However, some simultaneous equation pitfalls may occur; for example, if VMP, as well as supply of labor, is a function of climate, then estimates could well be biased. Hence, the usual hazards of single equation estimation may be amplified in the present analysis.

Although the model applies to the aggregate labor market of any urban area, it can be used for the analysis of specific occupation wage rates as well, since increased wages with size should occur for all occupations. Of course, city population will not be a scalar of labor quantity in a specific occupation; but this makes assuming that population size is exogenous even more reasonable.

Variables and Statistical Procedures

The application of the model can be described in terms of the selection of variables and statistical procedures employed.

A major problem in comparing wages is that of possible nonhomogeneity of the populations being compared. The Bureau of Labor Statistics (BLS) has developed wage data series on very precisely and narrowly defined occupations, greatly reducing this problem, and quite possibly eliminating it completely. For the present work, these BLS data were used to develop wage measures, the W of equation (1), organized into three samples covering different sets of occupations for specific years in the period 1966 to 1970. The Appendix contains detailed information on the specific sources of each sample. The number of observations, year covered, and occupations appearing in each sample are shown in the following list. Each occupation is referenced by its mnemonic code, which was employed in computer operations, and which is used here as reference abbreviation.

Sample I: 50 metropolitan areas, 1966 data, Building Trades. Metropolitan areas are Standard Metropolitan Statistical Areas (SMSAs).

MC—maintenance (nonunion) carpenter
ME—maintenance electrician
MP—maintenance painter

UC—union carpenter
UE—union electrician
UP—union painter

Sample II: 86 metropolitan areas, 1969 data, male plant occupations, and female office occupations.

JAN—janitor, male
LABR—laborer, male
MECH—auto mechanic, male
ACTAF—accounting clerk, class A, female
KP—keypunch operator, class B, female
STENO—stenographer, general, female
REC—switchboard operator-receptionist, female
TYP—typist, class B, female

Sample III: available wage data coverage for specific occupation ranged from 43 to 83 metropolitan areas, 1970 data, 7 office and professional occupations, and 4 plant occupations

CMPB—computer operator, class B
PROGB—computer programmer, business, class B
DRAFTA—draftsman, class A
DRAFTB—draftsman, class B
DRAFTC—draftsman, class C
ACTAM—accounting clerk, class A, male
BOY—office boy
NURSE—industrial nurse (registered), female
MACH—machinist
SHIP—shipping clerk
FLIFT—trucker, power forklift

In the raw data, wages for office occupations were on a weekly basis while those for plant occupations were on an hourly basis. In Sample II, the weekly figures were converted to an hourly basis using data on average hours in the work week for each occupation. Such data were not readily available for the office occupations of Sample III, so those wages remained on a weekly basis.

The wage observations were deflated by the monthly national consumer price index (CPI) with 1960 used as the base. This adjustment was carried out to make the three samples comparable, and, of more importance, to improve the comparability of the individual observations within a sample, since specific observations were obtained in different months of the fiscal year, running from July of the preceding calendar year through June of the listed calendar year, for example, July 1969 through June 1970, for Sample III. Finally, in most applications, a "standardized" wage was employed. This was obtained for each occupation by dividing by that occupation's average wage, and multiplying by

100, so that individual observations became percentages on a base of 100%, making occupational comparisons much more direct.

The 86 Standard Metropolitan Statistical Areas (SMSAs) of Sample II include the smaller set of SMSAs of the other samples as subsets. The larger size of Sample II, reflecting the greater availability of data on its occupations, made it the best sample of the three, *ex ante*, and it gave the best results, *ex post*.

There were three sets of nonclimate explanatory variables, the **Z** of equation (1), involving region, racial composition, and urban scale. Regional effects were measured by dummy variables covering the four major census regions: Northeast, North Central, South, and West. Two subregions of interest were included: the Confederacy and the Pacific Southwest. The first consists of states within the Confederacy during the Civil War, while the second consists of California and Arizona. Racial composition was measured by percent black and by an interaction term: percent black times the Confederacy dummy. (Dummy variables take on values of 0 or 1; thus, for the Confederacy dummy, if an observation falls within the Confederacy the variable takes on a value of 1, while for an observation outside the Confederacy, the value is zero). Urban scale was measured by (1) population size of SMSA in terms of the log of population, (2) the population density of the central city, and (3) the population density of the central city for cities with high density (over 10,000 per square mile), all other readings being set at zero. (The use of the log of population, rather than population, yields an improved statistical fit and does not really affect the central argument of the economic model, as illustrated in Figure 1.) Percent growth of population between 1960 and 1970 was introduced as an additional variable in Sample III, it being hypothesized that high growth places would tend to have wage rates somewhat above the average, with the reverse relation for low growth (or negative growth) places, reflecting presumed disequilibrium in process of correction.

Climatic variables comprising the **X** of equation (1) included summer temperature (July average, °F), winter temperature (January average, °F), and annual precipitation in inches. In addition, the following were also employed: the square of each of the temperature variables; the interaction of summer temperature and precipitation, measured by their product; and average wind velocity, in miles per hour. There was some experimentation with other climate measures including heating degree-days, inches of snow and sleet, annual number of days above 90°F, and annual number of days below 32°F. These variables essentially added nothing to explained variance, and were employed only in some subsidiary investigations.

Information on independent variables is presented in Table 1, which lists mnemonic code, full name of variable, unit, average value, and standard deviation for each variable. Averages and standard deviations are those that hold for the "full" set of observations of Sample II. (The one exception occurs for

Table 1. Information on Independent Variables

Category and Mnemonic Code	Full Name of Variable and Unit of Measure	Average Value[a]	Standard Deviation[a]
Region[b]			
NE	Northeast	0.256	0.436
NC	North Central	0.291	0.454
S	South	0.302	0.459
W	West	0.151	0.358
CONF	Confederacy	0.256	0.436
PAC	Pacific Southwest (Arizona, California)	0.070	0.255
%BLACK	Percent black	10.662	9.105
CB	Confederacy times percent black	5.294	10.118
LSPOP	Log SMSA population in 000	2.820	0.443
CDENS	Central city density in 000	6.004	4.424
HDENS	High central city density in 000 (>10)[c]	2.703	5.641
GROW	% growth in SMSA population, 1960–1970[d]	15.093	12.384
Climate			
STEMP	Summer temperature: July average, °F	75.530	5.183
WTEMP	Winter temperature: January average, °F	34.436	11.505
RAIN	Precipitation, annual average, inches	36.216	11.486
STEMP2/100	Square of STEMP divided by 100	57.317	7.893
WTEMP2/100	Square of WTEMP divided by 100	13.182	8.878
DAMP/100	STEMP times RAIN, divided by 100	27.488	9.455
WIND	Average wind velocity in miles per hour	9.624	1.696
DEGREE	Degree days (heating units) in 000	4.831	1.990
SNOW	Snowfall, seasonal average, inches	28.901	25.706
90+	Annual number of days over 90°F	32.105	31.085
32−	Annual number of days under 32°F	96.395	49.479

[a] Averages and standard deviations are for full set of 86 observations of Sample II.

[b] Dummy variable taking on values of 0 or 1.

[c] HDENS set at zero if CDENS < 10, hence, though value > 10 for nonzero cases, average is low because of large number of zero cases.

[d] Appears in Sample III only and ranges from 14.6 to 16.9 depending on SMSAs appearing. Value listed is for FLIFT equation with 83 out of 86 (maximum) observations.

179

percent growth in SMSA population, which appeared only in Sample III; its value in Table 1 is that for the equation with the largest number of observations in Sample III.)

The estimation of a "best fitting" equation for each occupation was carried out by developing equations containing only statistically significant variables, and hopefully, all such variables. The rationale for this procedure is that it eliminates "trivial" variables and reduces or eliminates problems caused by high correlation between independent variables that can cause low t ratios, peculiar coefficients, and large forecast errors. However, since the primary goal was development of a "best predictive" equation to gauge the effect of changes in climate, summer temperature, winter temperature, and precipitation were added to each "best fitting" equation, regardless of their significance. The rationale for this step is that it brings in the variables of basic concern, so that even if their effects are small, they are not disregarded. Of course, there is potential conflict with the previous rationale, but the prospective advantages were weighed more heavily than the disadvantages.

In developing the "best fitting" equation, the following rules were adopted:

1. A variable was retained in the equation if (a) it was statistically significant at the 10% level employing a two-tailed test or if (b) its retention caused another variable to become significant; in this case, it would be retained if its t ratio were above 1.0.

2. If there were two competing variables or two sets of competing variables, the variable or set selected gave the highest $\bar{R}^2$ (adjusted explained variance).

3. In moving to the final set, variables were eliminated in reverse order of t value, that is, those with lowest t values were dropped first. However, it is possible that a variable dropped at an early step could return to an equation after a number of other variables were dropped. To account for this possibility, eliminated variables were reintroduced at a later stage to see if they then became statistically significant.

Although these rules may have some arbitrary aspects, they appeared to be the best way of developing the specification of the equation. The first rule employed (basically, retaining variables with t at the 10% level) comes close to a rule specifying maximization of $\bar{R}^2$ (explained variance corrected for degrees of freedom). Dhrymes[2] has shown that fully maximizing $\bar{R}^2$ is equivalent to retaining variables with t ratios above 1.0, but that using this rule leads too frequently to the acceptance of false hypotheses.* Hence, the more conservative rule adopted here can be given some formal justification, since it will reduce the frequency of acceptance of false hypotheses.

---

* Proof that $\bar{R}^2$ is maximized when t values for all independent variables are greater than unity appears in Haitovsky,[3] while some additional discussion appears in Weiss.[15]

Fitted Equation Results

Sample I results for climate variables were not very encouraging, in terms of statistical significance. In the six individual equations for that sample, climate variables were often significant if regional dummy variables were excluded. However, when regional dummies were allowed to enter the equations, the climate variables almost always lost significance, being replaced in three equations by the Confederacy, in two by the South and in one by the West, though the additional amount of explained variance was typically on the order of .01 or .02. The Confederacy, for example, was well correlated with the three basic climate variables, STEMP, WTEMP, and RAIN, but hardly collinear, with an $\bar{R}^2$ of .68. Climate variables entered only one of the "best fitting" equations of Sample I, though at least one climate variable had a t ratio above 1.0 in the remaining five equations. Given the limited explanatory power of the climate variables in the individual equations of Sample I, an aggregate "best predictive" equation was developed directly and most economically by using an average standardized wage. Each observation on a given occupation's wage was divided by the average wage for the occupation, and then the relative wages were added across the six occupations for each SMSA. Dividing by 6 gave a construction industry standardized average wage, and this was used as dependent variable in a regression that brought in summer temperature, winter temperature and rain (STEMP, WTEMP, and RAIN) regardless of significance. Results appear in Table 2.

In Table 2 STEMP and WTEMP are of opposite sign; but the STEMP coefficient is a good deal larger than that of WTEMP; and the t ratio for the former variable is fairly high, though not significant. The RAIN coefficient is also negative. Hence, the Sample I results imply that a climate change involving a decline in temperature, or rainfall, or both, would yield an increased equilibrium, that is, would impose losses in real income that would have to be balanced by increases in money income. (This assumes that the amount of decline in temperature is the same in both summer and winter.) The evidence of the other samples indicate these conclusions must be tempered because relationships involve important nonlinearities.

In contrast to Sample I, climate variables performed well in the equations of Sample II in terms of statistical significance. Table 3 presents the coefficients, $\bar{R}^2$'s, and t ratios for the "best fitting" equations. Each column refers to a specific equation.

Summer temperature was the strongest of all the climate variables, entering all eight of the equations with negative sign; among the nonclimate variables, the log of SMSA population had the largest number of significant coefficients, appearing in seven of the eight equations, while high density* entered the eighth (and one other).

---

* This measure was city density times a dummy variable taking on 0 for density below 10,000 per square mile, and 1 for density above that level.

Table 2. Predictive Equation for Relative Wage, Construction Industry Workers (Six Occupations Combined), Sample I

Explanatory Variables	Mnemonic Code	Coefficient	t Ratio
Constant		100.4158	6.534
Region			
Northeast	NE	−6.2550	2.421
South	S	−5.3656	2.026
% Black	% BLACK	0.3543	1.974
Confederacy × % black	CB	−0.3300	2.151
Log SMSA population	LSPOP	6.7512	3.032
Summer temperature	STEMP	−0.2591	1.440
Winter temperature	WTEMP	0.0993	0.980
Precipitation	RAIN	−0.0947	0.922
R^2 = .5917			

Tabulating number of significant coefficients in Sample II by sign, we obtain:

Climate Variables	+	−
Summer temperature	0	8
Square of summer temperature	3	0
Winter temperature	4	0
Square of winter temperature	0	2
Precipitation	0	3
Interaction, precipitation and summer temperature	3	0
Wind	1	0

Nonclimate Variables	+	−
Log SMSA population	7	0
High density	2	0
Percent black	6	0
Interaction, Confederacy and percent black	0	6
South or Confederacy	0	5
Pacific Southwest	4	0
Northeast	0	2
West or Northcentral	2	0

There is considerable consistency here, and many of the sign patterns hold for all of the equations; thus, the square of summer temperature and wind generally had positive coefficients when brought into the equations in which they were not significant. At least one regional dummy entered every equation, and the pattern of signs was consistent across equations.

Explained variance adjusted for degrees of freedom ranged from .319 to .775, with more than half the variance explained in five of the cases. There is thus considerable explanatory power in the Sample II best fitting equations.

"Best predictive" equations were then obtained, and coefficients were summed by variable and divided by 8, the number of equations, to yield an aggregate predictive equation. That equation is presented at a later point with similar equations derived for the other samples.

Results for Sample III appear as Tables 4 and 5, the former covering office and professional occupations, and the latter covering plant occupations. These results appear to square rather well with those of Sample II, in terms both of coefficient sign and significance, though there is some tendency for the latter to be somewhat lower in Sample III.

Summer temperature, for example, was significant in six of 11 equations here, as opposed to all eight of Sample II; adjusted explained variance averaged .47 in Sample III, versus .58 in Sample II. The one exception to this pattern was wind velocity, which showed increased significance, appearing in six of the equations of Sample III.

The general result of lower significance levels may well reflect a smaller number of observations here; in some of the equations, the number was substantially smaller. This was probably a factor in the relatively poor performance of Sample I, as well. In particular, standard deviations for climate variables were somewhat smaller for Sample I than for Sample II, and with less variation in the independent variables, reduction in explanatory power can be expected. Some evidence supporting this inference is presented at a later point; that evidence, in turn, is relevant to future research strategy.

Tabulating the number of significant coefficients in Sample III, we find:

Climate Variables	+	−
Summer temperature	0	6
Winter temperature	3	1
Winter temperature squared	2	0
Precipitation	0	3
Interaction, precipitation and summer temperature	4	2
Wind velocity	6	0

Nonclimate Variables	+	−
Log SMSA population	9	0
Density and/or high density	4	0
Percent black	6	0
Interaction, Confederacy and percent black	0	6
South or Confederacy	0	7
Northeast	0	3
West	1	0

Table 3. Best Fitting Equations for Sample II

Independent Variable	Dependent Variable and Equation Results							
	JAN	LABR	MECH	ACTAF	KP	STENO	REC	TYP
	Coefficients							
Constant	186.423	657.584	571.024	370.933	101.613	98.448	116.225	108.488
NE	—	-11.158	-4.172	—	—	—	—	—
NC	18.476	—	—	—	—	—	—	—
S	—	-20.232	—	-5.708	-7.538	-6.106	—	—
W	—	—	6.354	—	—	—	—	—
CONF	—	—	-14.233	—	—	—	—	—
PAC	14.180	—	—	—	8.189	—	6.519	8.684
% BLACK	—	0.424	—	0.425	0.556	0.350	0.399	0.351
CB	—	-0.664	—	-0.424	-0.310	-0.239[a]	-0.270	-0.374
LSPOP	—	5.166	8.433	4.729	5.701	7.475	8.613	6.019
CDENS	0.646	—	—	—	—	—	—	—
HDENS	—	—	—	—	—	—	0.258	—
STEMP	-1.252	-12.903	-11.227	-6.798	-0.498	-0.389	-0.589	-0.369
STEMP2/100	—	7.017	6.107	3.944	—	—	—	—
WTEMP	—	0.678[a]	0.511	—	0.213[a]	0.269	—	—
WTEMP2/100	—	-1.195	-0.969	—	—	—	—	—
RAIN	—	-4.464	-4.754	-2.156	—	—	—	—
DAMP/100	—	5.904	6.337	2.931	—	—	—	—
WIND	—	—	—	—	1.039	—	—	—
$\bar{R}^2$	0.642	0.757	0.775	0.319	0.528	0.425	0.737	0.470

184

				F values			
Constant	12.641	3.426	4.985	2.716	7.051	6.917	13.362	9.266
NE	8.402	4.191	2.846	—	—	—	—	—
NC	—	—	—	—	—	—	—	—
S	—	5.436	—	1.961	2.347	2.051	—	—
W	—	—	2.271	—	—	—	—	—
CONF	—	—	6.003	—	—	—	—	—
PAC	3.645	—	—	—	1.963	—	3.243	3.242
% BLACK	—	1.947	—	2.496	3.303	2.106	3.467	2.429
CB	—	3.127	—	2.363	1.776	1.370^a	2.554	2.855
LSPOP	—	1.996	6.497	2.243	2.458	3.363	5.783	3.267
CDENS	—	—	—	—	—	—	—	—
HDENS	3.725	—	—	—	—	—	2.204	—
STEMP	6.533	2.706	3.933	1.984	2.635	2.082	5.255	2.444
STEMP2/100	—	2.354	3.423	1.810	—	—	—	—
WTEMP	—	1.374^a	1.746	—	1.605^a	2.805	—	—
WTEMP2/100	—	1.841	2.684	—	—	—	—	—
RAIN	—	3.032	5.550	2.146	—	—	—	—
DAMP/100	—	3.054	5.683	2.198	—	—	—	—
WIND	—	—	—	—	2.143	—	—	—

[a] Not statistically significant at 10% level, but appearance in equation causes another variable to become significant.

Table 4. Best Fitting Equations for Sample III, Office and Professional Occupations

Independent Variable	Dependent Variable and Equation Results						
	CMPB	PROGB	DRAFTA	DRAFTB	DRAFTC	ACTAM	BOY
				Coefficients			
Constant	113.874	89.896	71.681	64.008	113.533	156.322	150.843
NE	—	—	—	—	—	—	—
NC	—	—	—	—	—	—	—
S	—	—	-12.300	-6.737	—	—	—
W	—	—	—	10.075	—	—	—
CONF	—	-5.029	—	—	—	-9.194	—
% BLACK	—	—	0.762	0.679	0.328	0.216	0.551
CB	-0.250	—	—	-0.500	-0.462	—	-0.425
LSPOP	6.613	—	—	3.448	4.076	9.535	4.242
CDENS	—	0.601	—	—	—	-1.625	—
HDENS	—	—	—	—	—	0.517	—
GROW	0.147	0.097	—	—	—	—	0.173
STEMP	-0.456	—	—	—	-0.350	-1.289	-0.926
STEMP2/100	—	—	—	—	—	—	—
WTEMP	—	0.149	0.220	—	—	—	—
WTEMP2/100	—	—	—	—	—	—	—
RAIN	—	—	-0.206	0.370	—	-1.996	—
DAMP/100	—	—	3.062	1.211	—	2.878	—
WIND	—	—	—	—	—	1.542	—
$\bar{R}^2$	0.402	0.476	0.362	0.452	0.321	0.258	0.623

	t ratios						
Constant	7.374	35.254	9.397	8.611	7.823	4.073	12.591
NE	—	—	—	—	—	—	—
NC	—	—	—	—	—	—	—
S	—	—	4.109	2.631	—	—	—
W	—	—	—	3.205	—	2.301	—
CONF	—	2.659	—	—	—	—	—
% BLACK	—	—	5.641	4.265	1.974	1.401[a]	4.265
CB	2.347	—	—	3.016	3.078	—	3.608
LSPOP	3.413	—	—	1.730	1.853	3.204	2.271
CDENS	—	4.787	—	—	—	2.919	—
HDENS	—	—	—	—	—	1.342[a]	—
GROW	2.112	1.675	—	—	—	—	3.039
STEMP	2.341	—	—	—	1.948	2.813	6.280
STEMP2/100	—	—	—	—	—	—	—
WTEMP	—	1.833	2.245	—	—	—	—
WTEMP2/100	—	—	—	—	—	—	—
RAIN	—	—	—	—	—	1.612[a]	—
DAMP/100	—	—	1.801	2.838	—	1.779	—
WIND	—	—	4.026	2.517	—	2.644	—

[a] Each variable was significant when other two omitted; bringing in all three substantially improved $\bar{R}^2$.

Table 5. Best Fitting Equations for Sample III, Plant Occupations

Independent Variable	Dependent Variable and Equation Results			
	NURSE	MACH[a]	SHIP	FLIFT
	Coefficients			
Constant	57.169	66.597	248.256	174.895
NE	—	−9.120	−12.364	−12.221
NC	—	—	—	—
S	—	—	−16.429	−17.843
W	—	—	—	—
CONF	—	−15.833	—	—
% BLACK	0.376	—	—	—
CB	−0.473	—	—	−0.295
LSPOP	6.568	7.865	9.463	9.389
CDENS	—	—	—	—
HDENS	−0.404	—	—	—
GROW	—	−0.181	—	—
STEMP	—	—	−1.851	−1.474
STEMP2/100	—	—	—	—
WTEMP	0.345	—	−1.172	—
WTEMP2/100	—	0.476	1.690	—
RAIN	—	—	−2.555	−1.935
DAMP/100	−0.233	—	3.644	2.939
WIND	1.806	1.430	—	0.974
$\bar{R}^2$	0.453	0.575	0.541	0.684
	t ratios			
Constant	5.727	10.722	5.775	4.817
NE	—	4.836	3.968	4.669
NC	—	—	—	—
S	—	—	3.599	5.162
W	—	—	—	—
CONF	—	6.279	—	—
% BLACK	2.207	—	—	—
CB	2.632	—	—	1.922
LSPOP	2.402	3.898	3.859	4.655
CDENS	—	—	—	—
HDENS	2.243	—	—	—
GROW	—	2.525	—	—
STEMP	—	—	3.482	3.350
STEMP2/100	—	—	—	—
WTEMP	3.550	—	2.039	—
WTEMP2/100	—	3.357	2.356	—
RAIN	—	—	1.750	1.651
DAMP/100	1.853	—	1.936	1.927
WIND	3.495	2.789	—	1.921

[a] In an alternative specification with $\bar{R}^2 = 0.570$, significant variables included NC, CDENS, STEMP, STEMP2, RAIN, and DAMP, and excluded NE and GROW. In forming the best predictive equation, all other climate variables were employed, given the alternative specification results.

188

The pattern of plus and minus signs for temperature and precipitation shows a great deal of consistency between samples, the negative signs for STEMP and RAIN, and the positive sign for WTEMP of Sample I being repeated in almost all cases in Samples II and III. However, in the latter samples, the appearance of squared terms and DAMP, and their possessing of signs usually opposite those of their component variables, indicates that the effect of climate on wages is nonmonotonic, as well as nonlinear.

The evidence of all three samples furnishes good support for the hypothesis of increasing wage rates with urban scale, best measured by population size, which dominated density in explanatory power. It was expected that GROW, the measure of population growth, would have positive sign since faster growing areas should tend to have higher wages. There was some confirmation of this hypothesis, but it was not very strong, the variable entering only four equations of Sample III, with positive sign in three of the four, and its entry added little to explained variance. The variable was introduced at a late stage in the inquiry (the samples being numbered sequentially), and its limited explanatory power led to the decision not to bring it into the other samples, the cost of doing so seeming to outweigh any anticipated benefits.

In all three samples the South, Confederacy, and Northeast regions had negative coefficients when they entered the equation, while the West and Pacific always entered with positive sign, perhaps indicative of disequilibrium and direction of migration. The North Central region entered no equations in Sample III, and only one in Sample II, so that essentially its effect was neutral. Percent black and its interaction with the Confederacy each entered six equations of Sample III, the former always positive, and the latter always negative in sign, paralleling the results in the other samples. The sum of the two coefficients yields the net effect for percent black in the Confederacy. It was originally hypothesized that this effect would be negative, but typically, the two coefficients approximately cancel one another, so net effect is essentially nil. However, the net effect for laborers in Sample II in fact is negative, and this might well reflect a condition of relative oversupply of black workers in the Confederacy, and attendant outmigration. The positive coefficients for percent black outside the Confederacy might be interpreted as indicative of migration of blacks to areas of higher economic opportunity, evidenced in higher wages. They might also involve prejudice or externality effects.

The best predictive equations for Sample II were obtained in the usual fashion by bringing in basic climate variables when nonsignificant, and an aggregate equation was next obtained by averaging over the individual equations. The aggregate predictive equation for each of the three samples appears in Table 6.

The aggregate predictive equations were obtained by weighting each occupation equally. A more refined analysis might attach weights on the basis of each

Table 6. Aggregate Predictive Equations, by Sample

Independent Variable	Mnemonic Code	Sample I	Sample II	Sample III
Constant		100.4158	319.0014	168.3957
Region				
Northeast	NE	-6.2550	-1.9163	-2.9615
North Central	NC	—	2.3882	—
South	S	-5.3656	-4.9790	-4.9314
West	W	—	0.7943	0.6896
Confederacy	CONF	—	-1.7791	-2.6567
Pacific Southwest	PAC	—	5.9755	—
Percent black	% BLACK	0.3543	0.2985	0.2744
Confederacy × % Black	CB	-0.3300	-0.2839	-0.2503
Log SMSA population	LSPOP	6.7512	5.8881	5.6715
City density	CDENS	—	—	-0.0977
High city density	HDENS	—	0.1097	0.0121
Growth in SMSA population 1960–1970	GROW	—	—	0.0232
Summer temperature	STEMP	-0.2591	-5.3874	-1.5365
Square of summer temp/100	STEMP2/100	—	2.8768	0.4816
Winter temperature	WTEMP	0.0993	0.1694	-0.0640
Square of winter temp/100	WTEMP2/100	—	-0.2705	0.2168
Precipitation	RAIN	-0.0947	-1.4948	-1.0885
(Precip × summer temp)/100	DAMP/100	—	2.0497	1.3275
Wind velocity	WIND	—	0.1230	0.8001

occupation's share of the labor force, but that refinement did not seem reasonable at the present stage of analysis.

There is a fair degree of consistency in the values of nonclimate variables in Table 6. Thus, %BLACK has an average coefficient around 0.3, while that for CB is opposite in sign and a bit smaller in value. The log of population has a coefficient of around 6.0, indicating a 6% increase in wages for the same work for each increase in magnitude of population. Climate variable coefficients show somewhat less consistency, perhaps because of low significance in some of the underlying equations, as well as the differing role played by squared and interaction terms in the three samples. Yet, as noted earlier, a case can be made for a basic consistency of pattern, with negative signs for STEMP and RAIN, and positive signs for STEMP2, DAMP, and WIND. WTEMP and its square have signs opposite the summer temperature effects in Sample I and II, but parallel those effects in Sample III. However, the coefficients for winter temperature variables are considerably smaller than those for summer temperature, so that the impact on predictions of these differences should be marginal.

Predictions

The effects of a number of changes in basic climate variables were examined by applying the predictive equations of Table 6.

The present study was one of a number of investigations of the impact of climatic change,* and for that larger effort, a set of climatic changes by latitude was given. The following is the extreme or limiting case specified, organized in terms of latitudes applicable to the United States:

| | | Alternative Precipitation Changes (%) | | |
Latitude	Temperature Change, in °C	Increase	No Change	Decrease
20°—<30°	−1.5 to −2.0	+24%	0	−24%
30°—<40°	−2.0 to −2.5	+22.5%	0	−22.5%
40°—<50°	−2.5 to −3.0	+18%	0	−18%

The discrete changes listed were retained for precipitation, but temperature change was made continuous and converted to Fahrenheit by using the equations $\Delta C = .05\, L + .5$, which expresses the functional relation specified, and $\Delta F = 9/5\, \Delta C$, to convert from Centigrade to Fahrenheit, where ΔC and ΔF are the negative increments of temperature change in Centigrade and Fahrenheit, respectively, and L is latitude. Then, for each observation, i, in a specific

* The studies were carried out for the Climatic Impact Assessment Project, Monograph VI, supported by the U.S. Department of Transportation. See the acknowledgments of this paper.

sample, $\hat{W}_i$ and $\hat{W}_i{}^*$ were calculated, $\hat{W}_i$ being the standard predicted value of wages from the predictive equation, obtained by inserting average values of independent variables into that equation, and $\hat{W}_i{}^*$ the predicted value obtained by substituting climate variables that had been changed according to the specifications above. (The same temperature changes were applied to both summer and winter temperature variables.) Then $(\hat{W}_i{}^*/\hat{W}_i)$ was formed, expressing the ratio of predicted wages after the climate change to wages before the change. Finally, the average over the observations, i, was obtained for each case examined. Results are shown in Table 7.

In this table, predictions from the individual samples exhibit a fair amount of consistency. Those predictions were averaged by weighting by number of occupations in each sample, and the weighted average appears as the last column of Table 7. Employing that set, we find that the given temperature decrease, with no change in precipitation, is predicted as generating a 1.3% increase in money wages. In turn, the precipitation change implies an offsetting 0.6% change in wages; for the given temperature decrease, a concurrent decrease in precipitation implies a money wage increase of about 2% of the base wage, while an increase in precipitation leads to a money wage increase of only 0.7% of base income. The increase in money wages is interpreted as the compensatory payment that would be needed to keep real wages the same. If we assume the percentage change in wages applies to all income, then the climate change specified means a loss in real income of about 2%, in the worst case. If this were applied to national income, in 1973 the loss would amount to about $21 billion a year in total, or about $100 annually on a per capita basis.

Although the predictive equations are nonlinear, results do not seem markedly affected for scalar values of the original climate specification: thus, for climate change half those of the extreme case, wage increments were a bit less than half. In the case of both precipitation and temperature decrease, the weighted average of $(\hat{W}_i{}^*/\hat{W}_i)$ was 1.0083, corresponding to 42% of the original increment (.0083/.0196). A reversal of climate changes led to declines in equilibrium wages, and hence corresponded to increases in real income.

Table 7. Predicted Ratio of Wages after to Wages before Specified Temperature Decrease and Alternative Precipitation Changes

Precipitation Change	Sample I	Sample II	Sample III	Weighted Average
	(Wages After)/(Wages Before)			
Increase	0.9994	1.0181	1.0024	1.0067
No Change	1.0069	1.0191	1.0122	1.0132
Decrease	1.0145	1.0202	1.0220	1.0196

Table 8. Predicted Wage Ratios for Climate Change by Latitude Zones

	Specified Temp. Decrease & Alternative Precipitation Change		
Latitude Zone	Precipitation Increase	No Change in Precipitation	Precipitation Decrease
20°–<30°	1.0000	0.9999	0.9998
30°–<40°	1.0051	1.0094	1.0135
40°–<50°	1.0087	1.0176	1.0266

The nonlinearities do imply differing effects by latitude, and, of considerable interest, yield some implications for optimal climate conditions. The former result is illustrated by Table 8, which presents weighted average effects by latitude zone for the extreme temperature and climate changes specified earlier and which were averaged over all cases in Table 7. It can be seen from Table 8 that a temperature and precipitation decrease is neutral or even of minor benefit to the relatively hot and humid areas in the lowest latitude zone, but is costly in the other zones, and increases in cost with latitude. (This differential effect was more pronounced in the predictions from Sample II, which showed wage ratios of 1.012, 0.997, and 0.982 for the respective cases of Table 8 in the lowest latitude zone.)

These results suggest the existence of optimal climate conditions. Explicit inferences on optimality can be obtained by focusing on the three variables containing summer temperature, and rewriting the typical wage equation as:

$$W = a + bS + cS^2 + d(S)(P) \qquad (2)$$

where S is STEMP, P is RAIN, SP is DAMP, and the constant a contains all the other terms. Then setting the derivative, dW/dS, equal to zero and solving, we obtain:

$$S^* = \frac{-(b + dP)}{2c} \qquad (3)$$

where S^* is the solved value for STEMP. Typically, when the variables involved were significant, b was negative, while c and d were positive. Because c is positive, the second derivative, W, is typically positive, so wage is a minimum at S^*, corresponding to a maximum amount of benefits from STEMP at S^*; in short, S^* can be viewed as the optimal temperature. Further, as precipitation increases, S^* typically decreases. Hence, the optimum occurs at a higher temperature in a dry rather than in a wet climate, squaring with a number of common beliefs, for example, "it's not the heat, its the humidity."

For four equations in which all three variables of interest entered in the development of the predictive equations, the following results were obtained for S^*, with RAIN set at (1) its average value of 36.2 inches, (2) its average value plus one standard deviation, equal to 47.7 inches, and (3) its average value minus one standard deviation, equal to 24.7 inches:

Equation (Occupation)	"Optimal" Summer Temperature (S^*)		
	Rain = Average	Rain = Average + SD	Rain = Average − SD
LABR	76.71	71.90	81.57
MECH	73.14	67.17	79.11
ACTAG	72.80	68.46	77.14
MACH	74.24	71.28	77.19
Four equation average	74.22	69.70	78.75

For these equations, southern summer temperatures are well above the optimal level given the southern precipitation level. Using the machinist equation for the Confederacy, with average RAIN equal to 44.9, gave an S^* value of 71.98, as opposed to the actual average of 81.37.

Some additional investigation of optimality was carried out using the remaining equations of this study, by forcing in the climate variables of interest when they were not significant. Results tended to be in line with those reported above, but there was considerable "noise" in the results, reflecting often very low t ratios when the variables of interest were brought into the equations. Somewhat arbitrarily excluding some cases in which a wage maximum rather than wage minimum occurred, because of a sign change, these optimal temperatures by sample were calculated by averaging results for specific occupations:

Sample I   78.8°F   5 cases
Sample II   80.9°F   6 cases
Sample III   77.9°F   6 cases

These estimates hold for average precipitation, and will fall as precipitation increases.

These values are somewhat above the initial optimal estimate of 74°F. That initial estimate is within the range of a *comfort zone* of 73 to 77°F, given 20 to 60% relative humidity, as designated by the American Society of Heating,

Refrigerating, and Air Conditioning Engineers. An "ideal" or "effective" temperature has been related to relative humidity as follows:[9]

Humidity	Temperature
100%	68°F
50%	72.5°F
10%	75°F

These values seem fairly consistent with the range of temperature values presented above as a function of precipitation, using the four "best" equations for optimality inferences. However, it might be possible to rationalize the somewhat higher values obtained when all the equations are considered by arguing that the optimum should reflect not only physiological comfort, but all costs associated with STEMP and DAMP, including heating and air conditioning expenses, transportation, medical costs, and so on. It is possible that the economic optimum is shifted upward somewhat from the physiological optimum given an accounting for these other factors.

Some implications on effects of other basic climate variables are even more speculative, but may be noted in passing. The coefficient for precipitation depends not only on that for RAIN, but on that for DAMP, as well, given STEMP. Inserting the regional average for STEMP in individual equations in which both RAIN and DAMP were significant yielded a net coefficient for precipitation that tended to be positive for the Confederacy and negative for the West. These results square with the notion that the arid West would be better off with more rainfall, and that the South would be better off with less. Put more generally, the net coefficient for precipitation tends to change sign from negative to positive as STEMP increases.

In contrast to the strong performance of summer temperature, winter temperature contributed little explanatory power to the equations. But it is of some interest that the aggregate equation of Sample II in Table 6 implies that wages reach a maximum at a winter temperature around 32°F, indicating that at the freezing point, consumer welfare is at a low point, and real income increases to either side. If this result is taken seriously, some hypothetical explanations might be adduced, for example, winter transportation and medical costs may peak at that point. But this is highly speculative, at best.

In an extension of the analysis, the predicted wage ratio, $\hat{W}_i^*/\hat{W}_i$, was regressed on latitude; this relationship appeared to be linear, for the square of latitude, when brought into the equation, was never significant. But latitude was always significant. Table 9 exhibits results for regression of the wage ratio on latitude employing the Sample II predictive equation and given the extreme climate changes specified earlier. This kind of regression can be used to esti-

Table 9.   Regression of Predicted Wage Ratio on Latitude Using Sample II Results[a]

Climate Change: Temperature Decrease and Alternative Precipitation Changes	Constant	Coefficient of Latitude	$\bar{R}^2$
Precipitation increase	98.661	.0816	.092
No change in precipitation	92.651	.2400	.424
Precipitation decrease	86.639	.3990	.576

[a] Average of initial standardized wage is 100.

mate the wage ratio for any point, including points outside the United States, if one is willing to apply results outside the original range of the data. Thus in Table 9 the specified decreases in temperature and precipitation yield considerable benefits at the equator, estimated as roughly equal to a 13% rise in real income (via the 13% drop in money income).

As a final extension of the prediction effort, wind velocity was brought into the predictive equations, with a resultant coefficient of .719, that is, a 1 mph increase in wind velocity raises the standardized wage from 100.0 to 100.719. Intuitively, this seems reasonable because higher wind levels may cause higher heating costs or the need for stronger building construction. (The variable was not well correlated with STEMP or RAIN, but had a negative correlation with WTEMP of about .4, in line with the heating hypothesis, insofar as WIND is a causative factor in lower winter temperatures.) Given a 5% decline in wind velocity as an additional component of the extreme climate change considered earlier, and an average value of 9.624 mph, there is a standardized wage decline amounting to −.0035, offsetting somewhat the wage increases associated with the other climate changes. The offset is on the order of 20% of the wage increase (.0196) in the worst case shown in Table 7.

Climate and Region

In estimating climate impact by the wage relation of equation (1), there was some concern that effects attributed to climate might really involve region, instead. To avoid this difficulty, climate variables were forced to compete with regional variables. In fact, the competition may have been overly severe, because both the South and the Confederacy were allowed to compete, as were the West and the Pacific Southwest. In some cases, the Confederacy, but not the South, forced a climate variable out of an equation; in other cases, the South, but not the Confederacy, did the same thing. Each result was accepted. In the competition with the regional dummies, climate variables in Sample I did not do very well, in terms of statistical significance. But they did quite well in that competition in Samples II and III; often, both climate variables and regional variables were significant.

Turning the argument around, it was initially hypothesized that climate differences might explain all or most of the persistent North-South wage differential in the United States. The pattern of negative coefficients for STEMP and RAIN may appear to support the hypothesis, given higher than average values for those variables in the South, but the interpretation is shaken by the positive coefficients for STEMP2 and DAMP. The question was investigated in some depth by first regressing wages for each occupation on the South and LSPOP, only, and then on those variables plus all other significant variables from the best fitting equations, plus all climate variables (as listed in Table 3, for example). The ratio of Southern to Northern (non-Southern) wages was calculated for both cases using the coefficient for the South, as shown in the following paragraph.

Let $\hat{W} = \hat{A} + \hat{b}S$, where $\hat{W}$ is predicted wage, S is the dummy variable for the South, and $\hat{A}$ denotes the effect of all other variables, consisting of the sum of estimated coefficients times respective sample means. Then for $S = 0$ we obtain the Northern wage, and for $S = 1$ we obtain the Southern wage. In the first case $\hat{A}$ contains only the LSPOP effect, and in the second $\hat{A}$ contains the effects of all the other variables, as well. Results for this procedure are shown in Table 10.* Those results appear in terms of averages for all cases, and for a cross-classification of occupational type (blue-collar versus white-collar) and wage level (where an hourly wage of $2.50 or above is classified as high, and a wage under that figure is low).†

If the ratio of Southern to Northern wages increases after accounting for the effect of other variables, we can ascribe some of the initial differential to those other variables. A small reduction in differential is attributable to LSPOP, which is why that variable was included in the initial case. But the introduction of all other variables, including climate effects, had next to no effect on the differential, except in the case of low wage, blue-collar occupations. It is possible that climate affects working conditions more in those occupations than in others, with working conditions, as well as living conditions, affecting wages, but certainly this is not obvious. Outside evidence indicates the cost of living in the South is about 0.93 that in the North,[1] which exactly agrees with the all-

---

* In practice, calculations are simplified by noting that $\bar{W} = \hat{A} + \hat{b}\bar{S}$, where W is the sample mean and S is the mean for S, equal to the fraction of cases in the South relative to total sample size. (b typically has negative sign.) Then, $\bar{W} - b\bar{S}$ yields $\hat{A}$ or the Northern estimate, and $\bar{W} + b\,(1 - \bar{S})$ $= \hat{A} + \hat{b}$, or the Southern estimate. The ratio of Southern to Northern estimates is presented in Table 10.

† Wages are deflated to the 1960 price level. For the cases where wage data were on a weekly basis, wages were divided by an assumed work week of 37.5 hours.

Table 10.   Ratio of Wages in South to North Before and After Accounting for Climate Effects

Occupational Group[a]	Ratio of Wages in South to North		Number of Occupations
	Before Climate Effects	After Climate Effects	
Blue-collar			
Low wage	.835	.873	4
High wage	.930	.932	8
White-collar			
Low wage	.948	.950	6
High wage	.964	.955	7
All occupations	.929	.933	25

[a] Composition of occupation groups: blue-collar, low wage: JAN, LABR, SHIP, FLIFT; blue-collar, high wage: MECH, MACH, and all Sample I occupations; white-collar, low wage: ACTAF, KP, STENO, REC, TYP, BOY; white-collar, high wage: CMPB, PROGB, DRAFTA, DRAFTB, DRAFTC, ACTAM, NURSE.

occupation wage ratio in Table 10. However, the white-collar wage ratio is somewhat above this figure, while the low wage blue-collar figure remains well below it, even after accounting for climate effects. It is possible that some of the cost-of-living differential reflects an apparent "oversupply" of low skilled labor in the South, which depresses the wages of low skilled occupations. Hence, out-migration of low skilled persons and inmigration of higher skilled persons into the South may be one of the implications of Table 10.

In any event, it appears that the overlap of climate and regional effects is relatively small.*

Side Investigations

In the course of this study, several tangential analyses were carried out. Three of these side investigations are worth reporting in detail because they

---

* A similar conclusion can be derived from results in Izraeli.[7] He first uses laborers' wages as dependent variable, and then deflates each wage observation by the cost-of-living index for each urban area, thus removing the regional differential. In moving from the former to the latter case, the coefficient for winter temperature changes from −.50 to −.42, and the t ratio falls from 7.3 to 6.8. Hence eliminating the regional effect has only marginal impact on the climate effect. (Izraeli's use of only one climate variable limits the comparison, of course. Parenthetically, when I regressed laborer's wages of Sample II on winter temperature only, I obtained a coefficient of −.43, indicating some consistency between the two studies.)

yield insights of help in a critical appraisal of the central investigation, in turn useful in indicating the direction of future work that would build on the present effort. The three studies are listed under the headings of (1) effect of sample size, (2) pooling data into one sample, and (3) climate and urban scale relations.

Effect of Sample Size

It was suggested earlier that differences in sample size might explain the strong performance of the climate variables in Sample II relative to Sample I and to some of the cases of Sample III. This conjecture was investigated for two of the occupations of Sample III by comparing results obtained by using a sample of 79 observations to those obtained by using a subsample of only 42 observations. The first sample consisted of those SMSAs common to the two occupations with the largest number of observations, which were Draftsman, class B, with 82 observations, and Trucker, power forklift, with 83 observations; hence, coverage is close to that in the "full" set of 86 observations of Sample II. The 42-observation sample consisted of data from SMSAs common to the first set and to the 43 SMSAs appearing in the sample for Computer programmer, class B, the smallest set of observations in Sample III. Table 11 presents the estimates obtained for the two samples by regressing the standardized wages for the respective occupations on the South, the log of SMSA population, and the climate variables. In almost all cases lower values for coefficients, as well as for t ratios, occur in the smaller sample. The decline is especially pronounced for RAIN and DAMP, with coefficients in the smaller sample only a small fraction of those in the larger.

Beyond furnishing support for the conjecture explaining the relative performance of the samples, some additional inferences may be drawn from the evidence of Table 11.

1. Intuitively, it would seem reasonable to attach greatest weight to Sample II in drawing conclusions. In practice, this is not of much import for the present study, because predictions generated by the three samples were quite close (see Table 7). However, more substantial differences may well occur in other applications.
2. In future work, "large" samples seem mandatory. On the basis of the present results, a minimum sample size of around 70 observations seems necessary to avoid possible bias and to properly identify complex interactions, such as the temperature-precipitation interaction embodied in the variable DAMP. Considerable effort could well be devoted to development of even larger samples than that of Sample II.

Table 11.   Comparison of Results for Limited Set of
Independent Variables, Employing 79 versus 42
Observations, for Two Occupations in Sample III[a]

Independent Variable	Coefficients		t ratios	
	79 obs.	42 obs.	79 obs.	42 obs.
Case 1: Draftsman, Class B (DRAFT B)				
Constant	125.184	108.573	4.036	4.513
S	−8.465	−2.854	2.823	1.036
LSPOP	8.092	6.157	4.348	3.499
STEMP	−0.844	−0.488	2.168	1.699
RAIN	−1.806	−0.547	1.739	0.614
DAMP	2.564	0.761	1.900	0.655
WTEMP	0.692	0.648	1.621	1.229
WTEMP2/100	−0.829	−0.893	1.550	1.386
Case 2: Trucker, Power Forklift (FLIFT)				
Constant	195.151	169.128	5.074	5.016
S	−16.698	−11.333	4.691	3.067
LSPOP	10.769	6.241	4.633	2.610
STEMP	−1.713	−1.192	3.621	2.973
RAIN	−2.918	−0.466	2.414	0.388
DAMP	3.866	0.692	2.451	0.437
WIND	0.859	0.641	1.437	1.080

[a] $\bar{R}^2$: DRAFT B, 79 obs. = .335, 42 obs. = .389
$\bar{R}^2$: FLIFT,   79 obs. = .595, 42 obs. = .678

Pooling Data into One Sample

The use of standardized wages (wages as a percentage of the occupational average wage) permits the development of a larger sample by pooling of data on different occupations into one large sample. In practice, this was done for Sample II; thus, the 86 observations on the standardized wages of janitors were followed by the 86 observations on the standardized wages of laborers, and then by those on mechanics, and so on through the observations on typists, yielding 688 observations in all (86 observations per occupation times 8 occupations equals 688 observations). Of course, the data on the independent variables were simply repeated eight times, so that the sample coverage remains the original 86 locales. But the pooling does appear to improve statistical reliability by reducing the effects of variability specific to particular occupations. In the present application, pooling yielded some additional information by permitting (1) some analysis of the influence on wages of sex and skill, primarily in terms of the interaction of those factors with population size and the Southern regional effect; and (2) a deeper probing into effects attributable to the Southwest and the Pacific Coast.

For the first of these cases, the following variables were introduced:

SEX: a dummy variable, with male occupation set at 1, female at 0

AVGW: Average wage for specific occupation

SSEX: SEX times South (the dummy variable for the South)

SAVGW: Average wage times South

LSEX: SEX times LSPOP (log of SMSA population)

LAVGW: Average wage times LSPOP

SXAV: SEX times average wage

It should be noted that average wage is constant over the set of observations covering a particular occupation, for example, for auto mechanic it takes on a value of 2.98, and for keypunch operator, a value of 1.82.* It is interpreted here as primarily a measure of skill.

The consideration of the second case came about because of some internal evidence that Phoenix, Arizona had especially "good" climate, given the fitted equation results, and because it was conjectured that the original dummy variable used for the Pacific Southwest might not properly capture the effects of that region. In particular, it seemed possible that the climate of the Southwest and the Pacific Coast might differ enough from that for the rest of the country to lead to some regional effects on the coefficients of the climate variables. The availability of a larger sample allowed some experimentation along these lines, but the occurrence of only eight SMSAs (out of 86) in the region of interest limited the effort.

In the initial analysis, PAC was the dummy variable that took on a value of one for the SMSAs of the Pacific Southwest. In the extension here, it was augmented by these dummies:

PAC 01: Phoenix

PAC 02: Southern California, including Los Angeles, San Bernardino, San Diego and San Jose

PAC 03: San Francisco

PAC 04: Portland and Seattle

---

* There may be concern that use of this measure can violate the statistical assumption of independence of explanatory variable and disturbance term. Thus, a large positive error in an observation for a given occupation would tend to raise the average wage for the occupation somewhat, so that some nonzero correlation might then occur between the standardized wage (in effect measuring deviations from the average) and the average wage. But this seems a relatively trivial problem, for the correlation, if nonzero, should be very close to zero. In any event, AVGW had no discernable effect on the regression, since its t ratio was only 0.2.

In the initial formulation, PAC = PAC 01 + PAC 02 + PAC 03. San Francisco was treated as a special case here because of the possibility that perceived earthquake hazard increased the equilibrium wage in that SMSA. The revised formulation also included a combined dummy variable for the Pacific Coast:

$$COAST = PAC\ 02 + PAC\ 03 + PAC\ 04$$

Finally, two additional interaction variables were introduced:

CSTST: COAST times summer temperature
CSTWT: COAST times winter temperature

These last variables express the hypothesis that the Pacific Coast climate differs enough from the rest of the country to cause shifts in the coefficients of the climate variables. In some preliminary work, interactions of COAST and the other climate variables were introduced, but collinearity was so great, with correlations of the various interaction terms usually above .95, that it was clear only a few such terms could enter the equation, and the temperature variables were those then selected because they had the most explanatory power.

Table 12 exhibits the best fitting equation obtained using the pooled data and the additional explanatory variables. A number of those explanatory variables did not enter the equation, including SEX, AVGW, LSEX, PAC 03, PAC 04, PAC, and COAST.

Despite the entry of a number of the new explanatory variables, the pattern of results in Table 12 seems generally similar to the aggregate equation for Sample II in Table 6, obtained by summing individual predictive equations. Note that here the coefficients for the South and LSPOP depend on SEX and average wage. At the means for those variables, results in Table 12 become South: −7.552 and LSPOP: 5.012, versus the Table 6 results of South plus Confederacy: −6.758 and LSPOP: 5.888. The SEX and average wage results can be interpreted as follows (1) the Southern differential is greater the lower the skill, (2) the Southern differential is greater for male than for female occupations, (3) the population effect decreases with skill, and (4) male occupations pay more than female.

All of the climate variables enter the equation, usually with high t ratios, often well above the levels attained in the individual equations of Table 3. However, the explicit recognition of an effect for Phoenix causes the square of summer temperature to lose statistical significance, although its t value remains above 1.0 (at 1.2), and its appearance in the equation causes the Southern California dummy to attain statistical significance. The Pacific Coast temperature interactions are statistically significant, with coefficients opposite in sign to the basic temperature variables. In the case of summer temperature, this yields a relatively small change, with the STEMP coefficient equal to −4.916 for the

Table 12.   Regression Results Employing Pooled Data of Sample II (688 Observations)

Independent Variable	Code and Name	Coefficient	t ratio
	Constant	339.562	3.919
NE	Northeast	−1.925	2.006
S	South	−19.197	5.513
SSEX	(South)(Sex)	−16.372	11.897
SAVGW	(South)(Average Wage)	8.567	5.143
PAC 01	Phoenix, Arizona	28.552	4.544
PAC 02	Southern California	5.594	1.785
% Black	Percent Black	0.394	5.223
CB	(Confederacy)(% Black)	−0.451	6.335
LSPOP	Log SMSA Population	8.163	6.970
LAVGW	(LSPOP)(Average Wage)	−1.518	4.078
SXAV	(Sex)(Average Wage)	2.318	6.373
STEMP	Summer Temperature	−4.916	2.213
STEMP2/100	(STEMP) Squared/100	1.769	1.209[a]
WTEMP	Winter Temperature	0.398	2.169
WTEMP2/100	(WTEMP) Squared/100	−0.718	3.115
RAIN	Precipitation	−4.501	6.009
DAMP/100	(STEMP)(RAIN)	6.047	6.210
WIND	Wind Velocity	0.704	3.406
CSTST	(Pacific Coast)(STEMP)	0.500	2.653
CSTWT	(Pacific Coast)(WTEMP)	−0.635	2.072

$\bar{R}^2$ Adjusted explained variance .581

[a] Not statistically significant at 10% level, but appearance in equation causes another variable to become significant.

rest of the country and −4.416 for the Pacific Coast. In the case of winter temperature, a sign change occurs; thus, the WTEMP coefficient is 0.398 for the rest of the country and −0.237 for the Pacific Coast, implying that an increase in winter temperature is always beneficial in the latter region. Implications for optimal summer temperature for the rest of the country square well with previous results even though STEMP2 has fallen considerably in statistical significance; the optimal level here is estimated as 74.6°F on the basis of the Table 12 estimates and an average precipitation of 37.6 inches. The optimal for the Pacific Coast, with average precipitation of 20.3 inches, turns out to be 90°F, but this estimate is very sensitive to the level of precipitation, the optimal for Portland, Oregon being estimated as 60°, and that for San

Bernardino as over 100°. Intuitively, these results seem somewhat exaggerated. On the whole, however, the experiment with pooled data seems to have worked well enough to encourage further efforts along these lines.

Climate and Urban Scale Relations

The meteorological literature often considers the impact of cities on the climate.[8, 11] Hence, it seemed natural to relate climate variables to urban scale and geographic variables to investigate the relationship. However, we run into a problem of reverse causation, because, as argued in Figure 1, places with better climate will grow larger because of that quality characteristic. Hence, a simultaneous equation model may be necessary. But it is possible we may not go too far astray by the use of single equations, on the basis of results obtained by first relating climate to scale measures, and then by relating urban scale to climate. Different sets of variables are significant in the two cases, so that simultaneity may be less direct—and important—than initially perceived.

When climate variables were related to urban scale and geographic variables, best results were obtained for winter temperature and heating degree-days as dependent variables regressed on latitude, the West, the Confederacy, WIND, LSPOP (the log of population), and CDENS (central city density). Results are shown in Table 13, first excluding and then including CDENS. Exclusion of CDENS gave a population size effect that incorporated the effect of density, since density was fairly highly correlated with population size. Thus, variant 1 in Table 13 indicates that for each additional magnitude of population size, winter temperature will rise by 4.3°F; when we control for density, as in variant 2, the population size effect drops to 2.2° for each magnitude, that is, an SMSA of one million population will be 2.2° warmer than one of 100,000 population, if density is the same in the two places. WIND enters both equations, supporting the notion that increased wind causes lower temperatures and higher heating degree-days. The urban scale effects might be explained by the urban heat island, with both density and population size being factors in the generation of the heat island. Density is a somewhat stronger variable than population size for degree days, and considerably stronger for winter temperature, on the basis of the t ratios obtained.

Results for summer temperature were not nearly as clear-cut; LSPOP had a positive coefficient but its t ratio was only 1.06; CDENS had a negative coefficient, but there was an offset in large SMSAs by a positive coefficient for HDENS. Both density variables were significant. The fitted equation was:

$$\text{STEMP} = 108.2 - .86 \text{ LATITUDE} - 6.75 \text{ PAC} - 0.55 \text{ CDENS}$$
$$+ 0.36 \text{ HDENS} + 1.16 \text{ LSPOP}; \qquad \bar{R}^2 = .639$$

At the means, the population and density effects essentially are offsetting, and at high density the net effect of urban scale is positive, that is, STEMP

Table 13. Regressions of Winter Temperature and Degree-Days on Latitude, Region, Wind, and Urban Scale

Independent Variable	Dependent Variable and Equation Variant			
	WTEMP[b] (1)	WTEMP (2)	DEGREE[c] (3)	DEGREE (4)
	Coefficients			
CONSTANT	90.600	96.011	−5.143	−5.877
LATITUDE	−1.682	−1.721	0.303	0.308
WEST	9.026	9.990	−1.360	−1.491
CONFEDERACY	6.186	6.929	−1.122	−1.223
WIND[d]	−0.664	−0.702	0.075	0.080
LSPOP	4.292	2.231	−0.681	−0.401
CDENS	—[a]	0.319	—	−0.043
$\bar{R}^2$	0.886	0.892	.908	0.912
	t Ratios			
CONSTANT	11.825	12.269	4.320	4.795
LATITUDE	10.606	11.071	12.281	12.649
WEST	6.896	7.443	6.688	7.093
CONFEDERACY	3.688	4.160	4.304	4.687
WIND[d]	2.411	2.610	1.755	1.905
LSPOP	4.187	1.663	4.273	1.910
CDENS	——	2.301	——	1.992

[a] — Indicates variable not specified as being in equation
[b] WTEMP: Winter temperature (January average in °F)
[c] DEGREE: Heating degree days in thousands
[d] WIND: Wind velocity in miles per hour

increases with scale. However, the stronger relation for winter temperature than for summer temperature suggests that the heat island effect may be benign, on net, with winter benefits outweighing summer costs.

Other regressions showed snowfall to be negatively and significantly related to urban scale (density and population), but total precipitation to have a positive but not quite significant relationship. It can be hypothesized that increased precipitation is caused by increases in particulate air pollution with scale, but that the winter heat island effect outweighs the precipitation effect. These results square with the meteorological literature. However, wind velocity increases significantly with density (though it has no relation to population size), and this relation runs counter to what is reported in the literature. It is possible that intervening variables or reverse causation is involved; perhaps increased wind causes greater density.

Let us consider possible reverse causal relations in some detail, temporarily begging the question of simultaneity. When log of SMSA population is related to the climate and geographic variables, we find a number of statistically significant variables, as shown in Case 1 of Table 14. There appears to be considerable overlap with the results of Table 13. However, the Case 1 results are much improved by bringing in two new climate variables, labeled 90+ and 32−, which respectively measure annual number of days in which the temperature rises above 90°F or falls below 32°F. Neither these variables nor a measure of snowfall made any significant contribution to explanation in the case of the wage equations. In the case of population size, however, the first two of the variables replaced winter temperature, precipitation, and DAMP; eliminated latitude and the West; and added considerably to explained variance, as shown in Case 2 of Table 14. It appears, then, that days of extreme temperatures have an inhibiting effect on population size, with very

Table 14. Regressions of Log SMSA Population on Region and Climate Variables

Variable	Coefficient	t ratio
Case 1: Employing Initial Set of Climate Variables		
Constant	−1.9019	1.240
Latitude	0.0728	2.105
West	−0.4562	1.608[a]
Confederacy	−0.7109	4.066
WTEMP	0.0521	4.034
RAIN	−0.0681	1.874
DAMP	0.0812	1.743
WIND	0.0632	2.199
$\bar{R}^2 = .266$		
Case 2: Employing Extended Set of Climate Variables		
Constant	0.3134	0.285
Confederacy	−0.5617	4.218
STEMP	0.0390	2.489
WIND	0.0584	2.401
90+	−0.0065	2.521
32−	−0.0067	6.297
$\bar{R}^2 = .313$		

[a] Not statistically significant at 10% level, but appearance in equation causes another variable to become significant.

Table 15. Regression of City Density on Log SMSA Population, Region, and Climate Variables

Variable	Coefficient	t ratio
Constant	−32.892	3.273
LSPOP	5.992	9.018
Northeast	4.591	5.897
Latitude	0.177	1.671
STEMP	0.184	1.671
WIND	0.248	1.416[a]
90+	−0.044	2.521
SNOW	−0.031	1.691
$\bar{R}^2$ = .6658		

[a] Not statistically significant at 10% level, but appearance in equation causes another variable to become significant.

hot days and very cold days having essentially the same effect by way of approximately equal coefficients. An increase of one day of extreme temperature is predicted to reduce population by 1.5%.

Table 15 relates central city density to log population size and to climate and geography. It was assumed in the formulation that density is a function of population size, because with increases in population, land values are bid up everywhere within the urban area, causing more intensive use of land, and hence, higher density. The population size-density relation turns out to be quite strong, with a t ratio over 9.0. The only significant region was the Northeast, probably because the region's urban areas were settled earliest, with much of their development occurring prior to the "spread city" impact of the automobile. The remaining variables entering the equation were barely significant, save for days over 90° (90+). This variable and snowfall had negative coefficients, so that increases in these variables would be associated with declines in density.

Let us return to the simultaneity question by comparing Table 13 to the two following tables. In Table 13 we find degree-days is explained better than winter temperature, which in turn was explained better than days under 32°F (a side result not shown in the table) by inclusion of urban scale variables as independent variables. The summer temperature-population size relation was not significant, however. Finally, density performed better than did population in all these relations.

In Tables 14 and 15 winter temperature, per se, does not enter; but summer temperature does, as do the alternative measures 32−, 90+, and SNOW.

Hence, overlap of variables between equations is limited or absent, and this could be used to support the hypothesis that we have measured two distinct, and opposite, lines of causation by means of separate single equations. In any event, the simultaneity problem should not be of much concern in the primary study of this paper, because both urban scale and climate were treated as exogenous to wages. Of course, the scale-climate relation might best be examined by means of a simultaneous equation model, and that could be an important topic for future investigation.*

Conclusions

This section draws some conclusions and then expands upon earlier suggestions for future work. The following items are presented under conclusions.

1. Despite the limited success obtained with Sample I, there is good evidence from Samples II and III that climate variables are significant factors in wage relationships. Summer temperature was the strongest climate variable, but precipitation, wind, and winter temperature also were often statistically significant.

2. The wage-climate relation is nonlinear. The squared term for summer temperature and the interaction between summer temperature and precipitation, in particular, have the following implications:
   a. A summer temperature around 74°F is optimal, given average precipitation.
   b. Optimal temperature is inversely related to precipitation, rising as precipitation decreases and falling as precipitation increases.
   c. With a temperature increase, the net precipitation coefficient changes in sign from negative to positive, indicating precipitation shifts from being beneficial to being costly. This conforms to intuition: the arid West would do better with more precipitation, while the humid South would do better with less.

3. The wage-climate relations can be employed to predict the effect of changes in climate on welfare. Some predictions made here were based on a decrease in temperature and alternative changes in precipitation. The extreme case examined involved a decline in temperature of around 2.25°C, and a decline in precipitation of around 20%, on average. This implied an average reduction in real income of about 2%. However, the change in real income varied

---

* Meteorologists appear to be concerned about aspects of the simultaneity problem. Thus Helmut Landsburg, in noting climate differences between rural and urban places, expressed "nagging doubts" as to "whether the changes are due to urbanization or to natural differences in the places where cities are located." The point is noted in a discussion of the heat island effect ascribed to the new town of Columbia, Maryland, in "Columbia Makes it Hotter," *Washington Post,* February 3, 1975, B10.

with latitude, and the areas of the United States below 30 degrees of latitude, with high temperature and precipitation, are predicted to be somewhat better off, given the extreme case.

4. Regional and climate effects appear to be distinguishable. The hypothesis that the Southern wage differential reflects better climate had little support, receiving some substantiation only in the case of low wage, blue-collar occupations. It is possible that the Southern differential is in part explainable by an "oversupply" of labor in low-skill occupations.

5. There was some evidence that climate's effects on wages differed somewhat between the Pacific Coast and the remainder of the country.

6. There was evidence that increases in density and population size affect climate; this was interpreted as involving both the heat island effect and air pollution from particulates. The larger and the more dense the urban area, for given latitude, region, and wind velocity, the higher the winter temperature and annual precipitation, but the lower the snowfall, indicating that the heat island effect outweighs the pollution effect. There is some reason to believe the heat island effect is stronger in winter than in summer, which suggests it is beneficial, on balance. This conclusion is tempered by the likely mutual interdependence of climate and scale; however, different sets of climate variables seem important in the apparent reverse causal relations, and the lack of overlap involved may reduce, if not eliminate, the simultaneity problem.

Suggestions for future research can be based on discerned limitations or simplifying assumptions of the present study.

1. The relative lack of explanatory power for Sample I and some equations of Sample III can be explained by small sample size. Hence, larger sample size seems mandatory in future work, with a minimum size of around 70 observations.

2. The likely mutual interdependence of climate and scale suggests that a simultaneous equation approach should be considered in investigating their relationship.

3. The use of other climate variables merits investigation, including such measures as radiation, sunshine, fog, and measures of climate variability. Humidity might be employed in place of, or in addition to, precipitation. The correlation between the two variables was .716 (or $r^2 = .513$) using data on 192 U.S. weather stations, which is low enough to support the argument that experimentation with humidity could be worthwhile. The use of the *chill factor*, involving an interaction between winter temperature and wind velocity, also merits consideration. In the present study, however, some experiments with the product of the two variables yielded results that were quite discouraging. There is some evidence that the wind-chill relation

is nonlinear, however, with greatest chilling effect occurring when wind velocity increases from 5 to 10 mph[14]; hence, more complicated interaction terms might yield good results. The present study also experimented with three variables that measured extreme conditions: SNOW, inches of snow and sleet during the cold season; 90+, the number of days annually with temperatures above 90°F; and 32−, the number of days annually with temperatures below 32°F. All three turned out to be highly correlated with the basic climate variables, so it was not surprising that they had essentially no effect on results when they were introduced into the wage equations. They were of importance in the side investigation of scale and climate, but here they eliminated some of the basic climate variables. However, it is possible that a measure combining extreme conditions with variability would be useful in the wage equations, for the occurrence of unexpected extreme conditions can be costly. It can be hypothesized that short-period variability (sharp day-to-day changes) is a source of disutility that will cause a rise in money wages, but that longer-period variability (sharply defined seasons) may be seen as neutral or even of positive utility, and will lower wages.

4. Labor market conditions might yield some useful explanatory variables. The unemployment rate would appear to be a prime candidate under this heading, but there are indications that it need be handled cautiously. Unemployment should depress wages when it is a short-run, unexpected phenomenon, since declining demand for labor should have both quantity and price effects. However, if unemployment is a long-run, predictable phenomenon, for example, in a city whose industries are seasonal, or known to be especially subject to cyclic unemployment, then it is liable to be associated with increased wages, the differential representing a risk premium. This suggests the "raw" measure of unemployment may not perform well,* and that refined measures are necessary to distinguish between the two types of unemployment.

Other labor market conditions that might yield explanatory variables are extent of unionization; impact of the minimum wage—perhaps measured by combining samples on different occupations, and including a "low wage" dummy variable; and finally, the percent of black workers by specific occupation by SMSA rather than percent black for the entire SMSA population.

5. Other factors affecting the quality of life, such as earthquake hazard, or air quality, might be introduced into the wage equation. In his parallel study Izraeli[7] did in fact introduce air pollution, measured by sulfates, and obtained a positive coefficient that was significant at the 10% level. But cau-

---

* Izraeli[7] introduced unemployment into his equation and found it to have essentially a zero coefficient, with a t ratio of 0.5 in his money wage equation, and 0.1 in his deflated wage equation.

tion must be exercised here because of data limitations, the large number of potential variables (e.g., many air pollutants—particulates, carbon monoxide, nitrates) and the problems of collinearity. Air pollution levels and urban scale are highly correlated, for example.*

The foregoing illustrations involve sources of disutility, but certainly sources of utility (amenities) should be considered as well, for example, proximity to a large body of water for recreation.

6. Some of the problems of small sample size and many possible explanatory variables may be mitigated by combining standardized wages of several occupations into one sample, as carried out here in one of the side investigations.

7. Over time changes in technology or real income or tastes could well affect the parameters of the wage equation; for example, the money costs of air conditioning might be substituted for the psychic costs of bearing extreme summer temperatures—with a net decline in real costs, and hence in the wage differential, for hot places. This suggests as a line of inquiry the investigation of changes in the wage equation over time.

8. The present study did not consider possible population shifts in response to climatic shifts. Such shifts are likely to occur, and are rational if the cost of the move is less than the loss from staying in the same locale. Hence, the estimated cost of climate change presented in this paper could be an overstatement. However, the cost estimate here was based on changes in wages, and population shifts could also cause losses to capital-in-place, such induced costs tending to counter the cost reduction obtained from the population shifts. Further, calculation of costs at the margin can involve understatement, given a large enough shift. These points can be expanded upon as follows. If temperature and precipitation decline, a shift of population South can be anticipated, reducing the costs of climate change, or on occasion even generating benefits from them. (There are additional equilibrating factors, including changes in birthrates and in migration into or out of a country, which pose problems in evaluating change.) There will be transfers of property value along with population, with losses in capitalized value in the North tending to be balanced by gains in the South. However, aside from the direct costs of induced internal migration, some additional costs should occur. Some capital losses can be expected, because of inability to fully use up or move some capital fixed in place, such as urban infrastructure. Some land value losses in the North may not be balanced by gains in the South, if unique assets are involved. For example, Houston might not be

---

* See Hoch[5] for evidence on this relation. Izraeli's results show the air pollution t coefficient dropping substantially when population is included in the equation (from 3.3 to 1.8); yet the evidence in Ref. 5, p. 250 shows sulfates to be more strongly related to city density than to population size. This suggests that some of the effect now attributed to sulfates may really be attributable to density.

able to do as good a job as New York City if the former replaced the latter as our largest metropolis. Finally, wage differentials are calculated at the margin. If we move beyond that point, the costs of change may well increase. People who are specialized to a locale, including its climate, may lose a considerable amount of "rent" if there is a major change in their environment.

These second-order effects are important and merit considerable work, but are beyond the scope of the present study. (Their study would be important even if not tied to research on wage relationships.) It has been argued here that the study of wage relationships seems generally useful in addressing the question of the relation of climate to the quality of life. Perhaps that work could be extended to migration analysis that, in turn, might be of help in considering second-order impacts and induced migration flows.

Acknowledgments

This chapter is an extension of the paper "Wages, Climate and the Quality of Life," which was published in the *Journal of Environmental Economics and Management* during 1975. Aside from a few revisions, the first part of the chapter is identical to the journal article; however, from the section entitled "Side Investigations," to its conclusion, the chapter reports additional results not published elsewhere. The research for this chapter was partially supported by the U.S. Department of Transportation, CIAP Program, under contract DOT-OS-40032 with Dry Lands Research Institute at the University of California, Riverside. Such support does not imply endorsement, agreement with, or official acceptance of the research results contained herein. The extension of the work was supported by Resources for the Future. A detailed statement of the original set of research results appears in reference 10.
Judith Drake developed computer programs and carried out many of the computations for the work on this project. The manuscript was typed by Beverly Plater. I am indebted to Ralph d'Arge, Terry Ferrar, V. Kerry Smith, and George Tolley for useful comments that were applied in this work.

Appendix

Sources of Data

The data sources employed in this study can be classified into those for dependent and those for independent variables, and include the following documents:

Dependent Variables
Sample I: David, Lily Mary, and Kanninen, T. P., "Workers' Wages in Construction and Maintenance," *Monthly Labor Review,* January, 1968,

Table 1, "Straight-Time Average Hourly Earnings in Maintenance Work and Union Scales in Building Construction, 3 Trades in 50 Areas, 1965–66," p. 47.

Sample II: Bureau of Labor Statistics, *Area Wage Survey, Specific Metropolitan Area, 1968–1969,* Bulletin 1625-1 to 1625-90, Washington, 1970.

Sample III: Bureau of Labor Statistics, *Area Wage Surveys, Selected Metropolitan Areas 1969–70,* Bulletin 1660-91, 1971, Tables A-1, A-5, A-8. In Sample III, the number of observations, by occupation, were:

computer operator B	48	accountant A, male	74
computer programmer B	43	office boy	62
draftsman A	71	machinist	77
draftsman B	82	shipping clerk	74
draftsman C	76	trucker, power forklift	83
nurse	66		

Independent Variables

Statistics on independent variables were obtained from the following sources:

1. U.S. Bureau of the Census, *County and City Data Book, 1967,* Washington, 1967. Regional classification based on census regions: NE, NC, W, S, p. viii. Central city density for 1960 from Table 4. Temperature, precipitation, wind velocity, and heating degree-days, 65°F base, from Table 4 (STEMP, WTEMP, RAIN, WIND and DEGREE). The entries for Baltimore temperature in that source contained a typographic error and were corrected on the basis of information obtained from the Atmospheric Sciences Library, Silver Spring, Maryland.

2. U.S. Bureau of the Census, U.S. Census of Population: 1970, *Number of Inhabitants, Final Report,* PC(1)-A1, United States Summary, Washington, 1971. Central city density for 1970 from Table 30. Density for a given year between 1960 and 1970 (CDENS) obtained by interpolation.

3. U.S. Bureau of the Census, Census of Population and Housing: 1970, *General Demographic Trends for Metropolitan Areas, 1960 to 1970,* Final Report, PHC (2)-1, United States, Washington, 1971. SMSA population, 1970, from Table 11. U.S. Census Bureau, Census of Population: 1960, *Vol. I, Characteristics of the Population, Part A, Number of Inhabitants,* Washington, 1961. SMSA population, 1960, from Table 31. Population for given year obtained by interpolation from 1960 and 1970 data. Log of population (LSPOP) on base 10.

4. U.S. Bureau of the Census, U.S. Census of Population: 1970, *General Population Characteristics, U.S. Summary,* PC(1)-B1, Washington, 1972. Percent black by SMSA for 1970 from Table 67.

5. U.S. Bureau of the Census, *General Population Characteristics,* Final Report, PC(1)-iB, Individual State, Washington, 1961. Percent black by SMSA for 1960 from Table 21. Percent black for given year obtained by interpolation between 1960 and 1970 values.

6. *The Times Atlas of the World, Comprehensive Edition,* New York, 1967, for Latitude.

7. U.S. Environmental Data Service, *Climatological Data, National Summary, Vol. 20,* Asheville, N.C., 1967, 1968, and 1969 editions. Climate variables of inches of snowfall, number of days above 90°F and number of days below 32°F were obtained for each of those years by SMSA, and averaged to yield the variables employed (SNOW, 90+, 32–).

REFERENCES

1. Brackett, J. C., "New BLS Budgets Provide Yardsticks for Measuring Family Living Costs," *Monthly Labor Review* (April 1969), pp. 3–16.

2. Dhrymes, Phoebes J., "On the Game of Maximizing $\bar{R}^2$," *Australian Economic Papers,* Vol. 9, No. 14 (December 1970), pp. 177–185.

3. Haitovsky, Y., "A Note on the Maximization of $\bar{R}^2$," *The American Statistician* (February 1969), pp. 20–21.

4. Hoch, Irving, "Income and City Size," *Urban Studies* (October 1972), pp. 299–328.

5. Hoch, Irving, "Urban Scale and Environmental Quality," The Commission on Population Growth and the American Future, Research Reports Vol. III, Ronald Ridker Ed., (1972), pp. 235–284.

6. Hoch, Irving, "Inter-urban Differences in the Quality of Life," in J. G. Rothenberg and Ian G. Heggie, Eds., *Transport and the Urban Environment* Macmillan, London: 1974, pp. 54–98.

7. Izraeli, Oded, "Differentials in Nominal Wages and Prices Between Cities," Unpublished Ph.D. thesis, University of Chicago: 1973.

8. Landsberg, Helmut E., "Man-made Climatic Changes," *Science* (December 18, 1970), p. 1270.

9. Landsberg, Helmut E., *The Assessment of Human Bioclimate,* Technical Note No. 123, World Meteorological Organization, Geneva, Switzerland, 3 (1972), p. 9.

10. Panel on Economic and Social Measures of Biologic and Climatic Change, *CIAP Volume 6,* published by Institute for Defense Analyses for U.S. Department of Transportation Climatic Impact Assessment Project, DOT-TST-75-56 (1975), pp. 3–3 to 3–77.

11. Peterson, James T., *The Climate of Cities: A Survey of Recent Literature,* EPA (October 1969), 48 pp.

12. Resources for the Future, Inc., *Annual Report for 1972,* p. 70.

13. Smolensky, Eugene, "Industrial Location and Urban Growth" in Lance E. Davis et al., *American Economic Growth: An Economist's History of the United States,* Harper and Row, New York and Evanston, Illinois: 1972, p. 599.

14. Urdang, Lawrence, Ed., *The Official Associated Press Almanac,* Hammond Almanac, Maplewood, N.J., "Wind chill" table, citing National Oceanic and Atmospheric Administration, p. 281.

15. Weiss, Moshe "Letter to the Editor," *American Statistician* (June 1970) p. 20.

9

Climate Modification and Some Public Sector Considerations

PETER G. SASSONE
Department of Economics
Georgia Institute of Technology
Atlanta, Georgia

Climate may be defined as the characteristic atmospheric conditions of a region. In effect climate may be considered long term average weather, where weather encompasses precipitation, temperature, wind speed and direction, and ultraviolet radiation. One may speak of local, regional, continental, or global climate.

Changes in climate may occur as a result of three conceptually distinct processes.

1. Natural physical cycles. These are the cycles which have induced the periodic ice ages in "recent" earth history. Here the time scale is in thousands of years. We are not directly concerned with these cycles in this research, though our findings may have some "very" long term indirect relevance.

2. Human design. Controlling the weather is no longer a fantasy, even if it is not a widespread reality. Man is currently faced with two (related) long

This research was partially supported by the U. S. Department of Transportation, CIAP Program under contract DOT-OS-40032 with Dry Lands Research Institute at the University of California, Riverside. Such support does not imply endorsement, agreement with, or official acceptance of the research results contained herein.

217

term problems—energy and food. One of the key parameters of both these problems is the munificence of nature, that is, weather. The fortunate mildness of the 1973–1974 winter enabled the United States to fare well through a season that posed potentially great hardship to many. Thousands of Africans literally live or die each year depending on how the weather affects the harvests. Both energy and food problems, particularly the latter, are expected to grow chronically worse. Weather control may easily become a technological imperative. And, of course, long term weather control *is* climate control.

3. General economic growth. In this category we place all the actual or potential changes in climate induced as a *side effect* of (or *inadvertently* induced by) human economic (production and consumption) activity. Because climate is a long term concept, and because detailed and reliable data for whole areas has been collected for a relatively short period of time, scientists have yet to conclusively establish that climate has, in fact, been altered by economic activity. There is some dramatic evidence, however. LaPorte, Indiana has experienced significantly increased precipitation in recent years. It is hypothesized that this is the result of atmospheric particulates aiding condensation. La Porte is located downwind of the Gary, Indiana steel mills. Houston, Texas provides another example. Dr. Goldman (Associate Director of Research for the Institute for Storm Research) reported, in a recent seminar at the Georgia Institute of Technology, the results of a study of Houston's weather since the beginning of this century. Dr. Goldman felt the evidence supported the hypothesis that Houston's physical growth (tall buildings, etc.) influenced its climate. In addition to specific examples of history, most scientists would surely agree that economic activity is *capable* of affecting climate, whether or not it has yet. Environmentalists have long been warning that such activity may disturb whatever natural equilibrium we now enjoy.

The economics of the climate is at once an old and new topic. Agricultural economists of every bent recognize the importance of climate. So, too, do economists specializing in economic development or the location of economic activity. However, a whole new class of economic questions arises with the recognition that, in an age of widespread pollution and sophisticated technology, climate is better regarded as a variable than as a parameter in long term analyses of economic activity. Specifically, by adopting the public point of view, as compared with a more myopic production sector (e.g., agricultural) viewpoint, we pose the following questions:

• Who gains and who loses, and how much, when specific climate changes occur?

- Might some change in climate, for some region, whether inadvertent or designed, be a net benefit to that region?

- Is it possible that present atmospheric pollution is modifying climate in a beneficial direction?

These are very broad and very complex questions. They are posed merely to indicate the scope of the subject matter. We make no attempt to answer them in this chapter. Rather, as befits exploratory research, we must address the basic questions of methodology and the identification of impact areas. That is, where do we look for economic effects of climate modifications and how do we measure the benefits and costs once they're identified?

Our plan for this chapter is, first, to propose a methodology for investigating the economic effects of climate modification; second, to apply this methodology to a particular area of concern—the public sector; and finally, to make some estimates of public costs and benefits resulting from specific (assumed) climate changes.

Methodology

The rules governing our investigation are not novel and, in fact, they are quite conservative. However, it is imperative they be explicitly recognized.

First, only the *final* costs and benefits of climate modification are to be counted. This is in the same spirit, and for essentially the same purpose, as the economist's convention that only final goods and services be counted in the gross national product. This convention reduces the likelihood of double counting. For example, a colder climate entails greater space heating costs. *Ceteris paribus,* government might increase the size of welfare payments to compensate recipients for this additional expense. The observation would be an increase in government expenditure *and* an increase in personal expenditures. By focusing on final expenditures (the welfare recipients' outlay) we avoid incorrectly counting the same figures twice. In short, this procedure (final cost) applied to the government sector amounts to being wary of counting transfer payments as a real cost to society. As another example, consider the case of increased snowfall. To cover the increased costs of snow removal, road use taxes might be raised. Raw data would indicate increased consumer expenditure *and* increased government (highway maintenance) expenditure. Of course, only the latter should be counted.

Second, only *"first round"* costs are to be counted. The purpose of this point is unabashed simplification. We wish to avoid constructing and estimating a general equilibrium model for the economy. For example, a decline in temperature would induce increased fuel costs for space heating. This is the first round effect. The increased demand for petroleum products will raise their price, causing an increase in the price of plastics. This is a second round effect. The higher cost of plastic will raise the price of automobiles, for a third round effect. And so on. Adding to this potential round upon round morass are the cross effects between products. While the plastics situation is shifting the supply of autos to the left, the increasing price of gasoline is shifting the demand to the left. Consequently, one *could* observe a decrease in expenditures on autos and a concomitant decrease in air pollution. How to infer the social cost of the initial temperature decline, taking all effects into consideration, is an extremely difficult question. Seemingly, any proposal would be open to some challenge. Consequently, while we recognize that all effects should be counted, expendiency suggests we restrict consideration to initial, or first round, impacts.

Third, only effects on goods that are *climate inelastic* should be counted. Our intention is to focus on long run costs—those that will be incurred annually over the entire time horizon. Costs that occur because of lags in adjustment to a new climate will be ignored. In other words, we adopt a comparative statics point of view. There are long run annual costs under the given climate, and there will be long run annual costs under the new climate. The difference between these are the costs (or benefits) of the climate modification. But in order that this cost difference reflect a gain or loss in social welfare, the costs being compared must be for roughly the same quantity of the same good.

Some examples will clarify this. Consider the goods: in home temperature, snowfree streets, and public swimming pools.* Assume a climate change manifesting itself in the form of lower temperature. The demand for the first two goods is not likely to change much. People will still wish to keep their homes at the same level of warmth and their streets equally snow free. Only now the cost of providing these same goods has increased. This increase in cost is legitimately termed long run cost of climate modification. The same extra cost will be incurred every year. The case of the swimming pools is completely different, however. The temperature decline will reduce the resident's valuation of the pools, but their costs remain constant. This difference is a temporary, or transitional, social cost. For as the pools "wear out," not all will be replaced. Rather, more tennis courts might be built. After the residents have worked out their fiscal adjustment to the climate change, their new pattern of expenditure can give *no* indication of whether a net increase or decrease of social welfare has occurred.† This is not to say that no change in welfare actually occurs, only

---

* Assume the pools are financed by the sale of municipal bonds. Municipal tax revenues are used to retire a fraction of the debt each year. These annual "installments" are roughly equal to the aggregate value the residents inpute to a year's worth of municipal pools.

† Another way of looking at the problem of estimating climate induced increases in recreational expenditures is to consider that the individual's preference ordering over recreational goods is *con-*

that there is no ordinarily objective measurable manifestation of social cost of a climate change when climate elastic goods are involved.*

Fourth, both *inductive* and *deductive* reasoning should be employed in identifying climate sensitive cost and benefit areas. The overall objective of the economics of climate modification is a complete specification of all the cost and benefit impact areas along with associated estimates of the magnitudes of these effects. It is clear that neither induction or deduction alone is an efficient research method for this task. For pure induction would necessitate investigating every area empirically, while surely at least some could be deduced to be climate sensitive or nonsensitive. On the other hand, pure deduction would probably overlook many climate sensitive areas because our knowledge of every intricacy of all economic relations is far from complete.

The preceding four points obviously limit the range of investigation in pursuing the social cost of climate modification. Our rationale is that this approach solidly establishes a lower bound that enables us to confidently state that the social cost of some specific climate change is at least such and such. This type of hard figure is a necessary input to effective social policy formation.

We are under no illusion that these rules are complete in the sense of offering the solution to each methodological question that could arise in an investigation. Nor do we claim the rules are absolute. It is not difficult to conceive of research questions that demand alternative rules. Our intention is to state explicitly a methodology that we find quite useful, and suggest that others might also find it useful.

---

ditional on a particular climate. For example, let R represent miles of trout streams, X the number of outdoor tennis facilities, and T the temperature. A utility function may take the form

$$U = 2\sqrt{XR} + R(T - 65)$$

Consider two bundles of goods, X, R: $B_1 = (25, 30)$, $B_2 = (30, 25)$. Note that if $T = 64$, this individual would prefer B_1 to B_2, but at $T = 66$ he prefers B_2 to B_1.

Since most economists agree that utility must be interpreted ordinally, it is a difficult matter indeed to infer what monetary cost to assign to a drop in temperature to 64 from 66. The observed increases or decreases in expenditures on X and R would provide little basis from which to infer social welfare costs.

* This point in our methodology—that we limit our attention to climate inelastic goods—is certainly not a binding role on future research. It results from the dual considerations of expedience and conservatism, the latter in that the substitution among goods as climate changes opens up a Pandora's box of difficulties. Nonetheless, situations may arise wherein the nature of substitution effects are clear *a priori,* and traditional welfare economics analysis may be applied. In fact, if not for a lack of suitable data, that approach is conceptually proper for the estimation of the incremental highway maintenance costs discussed later in this chapter.

Public Sector Costs of Climate Modification

The U.S. public sector has attained an overwhelming size and complexity. It, alone, accounts for more expenditure than the vast majority of the world's economies. Here, we focus on only two aspects of the public sector: state highway snow removal costs and local functional expenditures. The former is clearly suggested by deductive reasoning, whereas the latter is fertile ground for an inductive approach.

Local Expenditures

Searching the literature, we can find no previous attempts to systematically relate local public expenditures to climate, although Brazer's[1] pioneering work does allude to the problem. Regression analysis on cross-section data, the usual technique in attempts to explain local expenditures, is adopted here. Our sample consists of 196 cities for which meteorological, demographic, and economic data are available.[4, 5] The dependent variables include total municipal expenditures, noncapital municipal expenditures, and expenditures on the functional categories of education, welfare, police and fire services, sanitation and sewage, and recreation. It was felt that even if the first two variables showed no climate sensitivity, it might be due to self-cancelling climate effects on their components—the functional categories. Thus, the reason for the inclusion of this latter class of variables.

Of the independent variables, 25 economic-demographic (census) variables were collected. Due to expected multicollinearity, however, many could not be used in the regression equations. And of those truly "independent" variables, relatively few explained most of the variance in the dependent variables. Twenty-one climate variables were collected. Again, a good deal of multicolinearity was experienced. For example, the following variables were found to be related: normal July daily maximum temperature, normal January daily maximum temperature, normal annual temperature, normal seasonal degree-days, mean seasonal snow, annual mean number of days temperature exceeds 90°F, and annual mean number of days of snow 1 in. or more.

Tables 1, 2, and 3 report the results of the regression analysis. Note, in Table 1, that climatic effects are not significant at the 10% level in explaining *total* municipal expenditures. For two reasons, this result is not surprising. First, as mentioned above, climatic effects manifesting themselves in the functional categories might tend to self-cancel in aggregation. Second, climatic effects, if present, are likely to be small. Hence, aggregation might tend to swamp them in random noise and unexplained effects. Tables 2 and 3 contain the results of the investigation of climatic effects on functional expenditure cate-

Table 1.   Regression[a] Results Explaining Total City Expenditures (1970)

Dependent Variable	Total Municipal Expenditures (Millions)			
Independent Variables	Dimension	Coefficient	STD. Error	t Statistic Significance
Constant	—	22.96191	—	—
Area	Square Miles	−0.42759	0.14428	.995
Population change (1960–70)	% °F Inches	0.41743	0.44490	.80
Mean annual temperature		−0.98801	1.90107	.60
Mean annual precipitation		0.98414	0.81840	.80
Mean number of days temperature reaches 90°F	—	0.30168	0.44596	.70
Increasing function of population[b]	—	0.34395	0.00839	.995
Percent of dwelling units in single family homes	—	−0.00002	0.00000	.995
Number of observations: 196				
F Ratio: 299.034				
R^2: .92				

[a] Using ordinary least squares with a linear regression equation.
[b] It was found that raising population to the 1.1 power gives a slightly better fit. This accounts for diseconomies of scale in large cities.

gories. Separate tables are used since the category of recreation expenditures was not reported in the 1970 census, so a sample of 46 cities from the 1960 census was used for that item. Note that there are significant climatic effects on education, sanitation and sewage, and recreation. At present, we can only hypothesize about the causal relations involved. With respect to education, we find a positive relation between municipal expenditures and annual precipitation. This can be attributed to differences in construction costs. Less precipitation would allow greater use of open air corridors between buildings and greater use of open air stairwells for multilevel buildings. Presumably, this is a less costly construction technique. The relation between sanitation and sewage expenditures and number of rainfalls over 0.01 in. can also be attributed to capital costs, storm sewers in particular. Less rainfall means a decreased need for such sewers. In fact, many cities have no storm sewer facilities at all. This is a saving of capital and maintenance costs, and would account for the relation discovered.

The results in Table 3 certainly conform to expectations. Cold climates mean less publicly sponsored outdoor recreation—such as parks, playgrounds, pools,

Table 2. Regression Results for Selected Municipal Functional Expenditures (Dependent Variables in Thousands of Dollars. 1970 Data)

Dependent Variables	Constant	X_1 Coefficient[a]	t Significance[b]	X_2 Coefficient[a]	t Significance[b]	X_3 Coefficient[a]	t Significance[b]	X_4 Coefficient[a]	t Significance[b]	X_5 Coefficient[a]	t Significance[b]	X_6 Coefficient[a]	t Significance[b]
Education	−377.45	8.81770	.995	−0.064	.995	−8.97277	.975	—	—	29.77084	.90	—	—
Welfare	1152.54	8.75965	.995	−0.059	.995	−11.50155	.99	−0.24204	.95	—	—	—	—
Police and Fire depts	−145.70	3.59557	.995	−0.007	.995	−3.12418	.995	0.05646	.995	—	—	—	—
Sanitation and Sewage	−183.41	1.20453	.995	−0.002	.995	—	—	0.01426	.90	—	—	2.27685	.995

[a] X_1 is a population variable (thousands of population raised to the 1.1 power)
X_2 is the number of single family homes
X_3 land area of municipality, in square miles
X_4 is population density = X_1/X_3
X_5 is annual precipitation in inches
X_6 is number of rainfalls over 0.01 in.
[b] Only significance levels of .90 or better are reported.

Table 3.   Regression[a] Results for Municipal Reaction Expenditures (1960)[b]

Dependent Variable	Municipal Recreation Expenditures (Thousands)			
Independent Variables	Dimension	Coefficient	Std. Error	t statistic Significance
Constant	Thousands of dollars	112.44127	—	—
Population	Individuals	0.02442	0.00703	.995
Aggregate income	Millions	−5.05429	2.69191	.95
Mean number of days temperature drops to 32°F	—	−6.49502	4.79252	.90
Number of observations: 46				
F Ratio: 107.768				
R^2: .89				

[a] Using ordinary least squares with a linear regression equation.
[b] 1970 Data not available for this expenditure category.

athletic fields, tennis and basketball courts.* The important variable here is "number of days temperature drops to 32°." This indicates our interpretation of causality has some appeal. For there is no relation of recreation expenditures to mean annual temperature, thus we are not forced to hypothesize that cooler weather diminishes that type of public expenditure, since cooler weather (e.g., from 70 to 60°F) could promote recreation.

What can we conclude about the local public costs of climate modification? As discussed above, it would not be proper to draw any conclusions about true long term social costs on the basis of the recreation results. Here there are a great many substitute goods in the private sector, so inferences about gains or losses of social welfare from public expenditure data cannot be justified. To reiterate, there may *in fact* be losses or gains, but they cannot be inferred from expenditure data.

On the other hand, the education and sewage costs would seem to reflect demands for climate inelastic goods: comfortable movement around the educational facility and nonflooded streets and homes. These would likewise qualify as first round and final costs. Therefore, we can conclude that climate (in

* Casual empiricism suggests that public recreational expenditures are mostly for "outdoor" goods. This is no accident—it's precisely what the theory of public expenditure would predict: where exclusion is not a viable alternative—as in many outdoor projects—the good, if provided at all, must be provided by the public sector. In addition, outdoor recreation is also publicly provided because it is considered a "merit good." An elaboration of the theory can now be found in recent texts such as.[2, 3]

particular precipitation) modification *will* affect long term social costs in public sector education and sewage; furthermore, the magnitudes of such costs or benefits are given in Table 2.

The reader may object to the attempt to find a linear relation between climatic variables and expenditures. Numerous hypotheses may be advanced implying other forms of relations, chief among these being the U and inverted U forms. These possibilities were tested for the mean annual temperature variable. The sample of 196 cities was partitioned into two sets: those over and those under a mean annual temperature of 60°F. The dependent variable is per capita noncapital municipal expenditures. Results are displayed in Tables 4 and 5. Temperature turned out to be a significant variable in the under 60° group, but does not appear to affect expenditure in the over 60° group. Figure 1 summarizes this result. Of course, analysis of the functional expenditure categories and of other climatic variables may well prove to afford greater insight into the climate modification problem. For the present, we are content to simply point out that this appears to be fertile ground for further investigation.

State Highway Maintenance Expenditures

The inclusion of this class of expenditures is based on deductive logic. Increased or decreased snowfall will clearly increase or decrease highway maintenance, specifically snow removal, costs. It represents *final, first round* costs for a service that may reasonably be supposed to be climate inelastic, at

Table 4. Regression[a] Results for Cities Whose Mean Annual Temperature is Less Than 60°F (1970)

Dependent Variable	Per Capita Noncapital Municipal Expenditures (Dollars)			
Independent variables	Dimension	Coefficient	Std. Error	t Statistic Significance
Constant	Dollars	107.12490	—	—
Population density	Individuals per square mile	0.01005	0.00320	.995
Percent of dwelling units in single family homes	—	−3.69267	0.78574	.995
Mean annual temperature	F	3.84264	2.19647	.95
Mean annual precipitation	Inches	1.45522	0.99467	.90
Number of observations: 123				
F Ratio: 25.002				
$\bar{R}^2$: .52				

[a] Using ordinary least squares and with a linear regression equation.

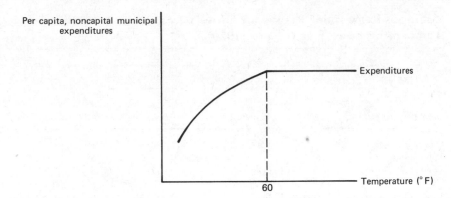

Figure 1. The relation between mean annual temperature and per capita noncapital municipal expenditures.

least for marginal changes. Our approach is to construct a model explaining snow removal costs; estimate the parameters using available data; and test its explanatory power; that is, test the model as a hypothesis to confirm or reject our deductive logic.*

Four elements enter into the determination of annual costs of road maintenance. These are the number of snowfalls, the number of miles of the road system, the average intensity of maintenance, and the cost per mile of maintenance. The penultimate factor demands some clarification. Everyone who has lived in the northern half of the United States has probably, from time to time, found certain roads or streets still uncleared days after a snowstorm. While locally frustrating, this results from a deliberate priority system of the relevant (municipal or state) highway department. The priority accorded a road segment depends on, among other things, the availability of alternate

---

* This approach, the estimation of a snow-removal cost function, is not the only conceptual approach to the problem. Alternately, one could attempt to estimate the demand for snow removal and use a consumers' surplus approach to estimating welfare losses (or gains) due to changes in the "price" of snow-removal. The problem with this approach is the lack of relevant data. This becomes evident when one realizes that consumers' demand is not for snow removal *per se,* but for clear streets. Surely there is an inverse relation between the price of keeping streets cleared of snow and the level of clearedness that consumers choose. The difficulty is in getting actual quantitative measures of the price per unit of clearedness and the units of clearedness that a cross section of municipalities have faced and chose, respectively, for some recent year. To the best of our knowledge, such data is not available. The approach adopted here takes advantage of the fortunate circumstance that the demand for cleared streets is very likely inelastic over the relevant range. Thus, whatever the number of units of clearedness currently being consumed, we assume that amount will not change much in response to small changes in snowfall. Thus, the greatest part of the costs of an increase in snowfall is in the increase in maintenance costs, not in a welfare loss associated with less units of clearedness being available.

Table 5. Regression[a] Results for Cities Whose Mean Annual Temperature is 60°F or Greater (1970)

Dependent Variable	Per Capita Noncapital Municipal Expenditures (Dollars)			
Independent Variable	Dimension	Coefficient	Std. Error	t Statistic Significance
Constant	Dollars	232.19753	—	—
Population	Individuals	−0.00001	0.00001	.80
Population density	Individuals/ sq. mile	−0.00366	0.00362	.80
Median family income	Dollars	0.00495	0.00306	.90
Percent of dwelling units in single family homes	—	−2.00424	0.54623	.995
Mean annual precipitation	Inches	−0.33804	0.25273	.90
Number of observations: 73				
F Ratio: 5.243				
$\bar{R}^2$: .28				

[a] Using ordinary least squares and with a linear regression equation.

routes, the impact on the local economy (e.g., business districts are almost always maintained), health and safety considerations (access to hospitals, police, and fire department routes), and whether influential citizens live along that segment. Quantitatively, intensity can be considered the proportion of all road mileage that is actually maintained during and following a typical snowfall (or sleet).

Before explicitly setting out our model, some notation is needed.

N: number of snow (or sleet) falls
M: total road mileage of the area under consideration
I: intensity of road maintenance
A: depth of average snowfall
P: population of the area
V: value added by manufacture in the area
W: number of employees in wholesale and retail trade in the area
S: total annual snowfall (in inches) in the area
TC: total cost of road maintenance for the area
C: cost per mile of maintenance
$a, b, \alpha, \beta, \sigma, \partial$: parameters of the model

The basic relation in the model is

$$TC = N \times M \times I \times C \tag{1}$$

where

$$I = f(P, M, V, W) \qquad (2)$$

and

$$C = g(A) \qquad (3)$$

The precise form of equation (2) is not evident on any *a priori* grounds. The relation should be such that I increases with increases in P, V, and W, since greater population and greater industrial and commercial activity should increase the proportion of roads maintained. On the other hand, as M increases the number of alternate routes tends to increase. Thus, we assume as M increases I decreases. In keeping with the multiplicative form of equation (1), we can rewrite equation (2) as

$$I = aP^{\alpha} M^{\beta} V^{\sigma} W^{\partial} \qquad (4)$$

where a, α, σ, $\partial > 0$ and $\beta < 0$.

The simplest form for equation (3) is a linear homogeneous relation—when the depth of snowfall doubles, costs per road mile also double. An objection might be raised here that the process of road maintenance entails large fixed costs and that increases in snowfall represent only increases in variable cost. This would be a valid objection if our purpose were to estimate costs for, say, a forecasted harsh winter. However, our interest is in the effects of climatic (long term) change, hence we adopt the usual assumption that in the long run all costs are variable. Equation (3) may be specified as

$$C = bA, \qquad b > 0 \qquad (5)$$

Substituting equations (4) and (5) into equation (1) yields the final form of our model:

$$\text{TC} = abSM^{1+\beta} P^{\alpha} V^{\sigma} W^{\partial} \qquad (6)$$

Equation (6) is readily converted into log linear form, which suggests estimation by way of ordinary least squares. Data on TC and M are drawn from *Highway Statistics, 1970.** Maintenance costs are most reliably reported for state controlled roads and streets, which is why we decided to base our estimation on that subset of the United States road network.† As a further limiting factor, we decided to exclude southern states from the data set. Casual empiricism suggests these states are typically unprepared for whatever sporadic snow

---

* A U.S. Department of Transportation/Federal Highway Administration publication.
† As of 1970 the U.S. road and street network totaled 3,730,082 mi, of which 781,105 mi were state controlled, 2,761,281 were locally controlled, and 187,696 were under federal control.

Table 6.   Correlation Matrix

	S	M	V	W	P
S	1	−.241	.080	.052	−.130
M		1	.335	.354	.453
V			1	.961	.867
W				1	.873
P					1

or sleet that they receive, hence their reported costs may reflect a different road maintenance production function.*

Observations on S are long-term averages, and P, V, and W, like TC and M, are 1970 figures.† There is a potential difficulty arising from the lack of correspondence between TC and S. TC is the vector (33 observations) of costs actually incurred in 1970, while S is average, or expected, snowfall, not the actual 1970 readings. This disparity looms less serious when one recognizes that both capital and labor are purchased or hired, not on the date of the snowfall, but prior to the winter season. Thus, to a large extent, actual costs (wages, depreciation, etc.) reflect *expected* conditions more so than the *actual* conditions of any one year. We feel, therefore, that the long term average S is the appropriate independent variable.

Glancing again at the independent variables in equation (4), one might easily suspect that multicollinearity is present among P, V, and W. Table 6 verifies such suspicion. While multicollinearity will not bias the estimate in *OLS,* the importance of each variable will be obscured. We remove V and W from the estimating equation by using the relations $V = a_1 P^{\alpha,1}$ and $W = a_2 P^{\alpha,2}$. Equation (7) becomes

$$TC = (aba_1a_2) \, SM^{1+\beta} \, P^{\alpha + \alpha,1 + \alpha,2} \tag{7}$$

---

* Thirty-three states remain in the data set: Colorado, Connecticut, Delaware, Idaho, Illinois, Indiana, Iowa, Kansas, Kentucky, Maine, Maryland, Massachusetts, Michigan, Minnesota, Missouri, Montana, Nebraska, Nevada, New Hampshire, New Jersey, New Mexico, New York, North Dakota, Ohio, Pennsylvania, Rhode Island, South Dakota, Utah, Vermont, Virginia, West Virginia, Wisconsin, Wyoming. Alaska is not included because its economy appears to be significantly different from those of the lower 48.

† P, V, W are from the *Statistical Abstract of the United States, 1972*. S is reported in United States Department of Commerce, *Climatological Data National Summary,* Annual 1966, Vol. 17, no. 13. Readings are taken at various observation sites. We are using the average value for all sites in a state.

which, for estimation purposes, is more conveniently written as

$$TC = \beta_0 \, S^{\beta_1} \, M^{\beta_2} \, P^{\beta_3} \tag{8}$$

$$\left. \begin{array}{r} \beta_0 > 0 \\ \beta_1 = 1 \\ 0 < \beta_2 < 1 \\ \beta_3 > 0 \end{array} \right\} \tag{9}$$

Equation (8) and Restrictions (9) constitute our hypothesis on the factors affecting highway maintenance costs. A high level of significance for β_1, β_2, β_3 *and* satisfaction of (9) will cause us to accept the hypothesis. Failure on one or more counts will be interpreted as reason for a strategic retreat to the modeling stage.

The coefficients were estimated using the data described above. Table 7 presents the results.

The reader will note that all the criteria for acceptance of the hypothesis are met. In particular, note that $\beta_1 = 1.07$ is not significantly different from unity at the .90 level. The estimated equation, $TC = 0.03 \, S^{1.07} \, M^{0.22} \, P^{0.75}$, has some interesting properties. Not only is it approximately homogeneous of degree one in S, as was hypothesized, but it is also approximately homogeneous of degree one in M and P (jointly). Interpretation is straightforward. *Ceteris paribus,* doubling snowfall doubles the cost of road maintenance; and, *ceteris paribus,* doubling *both* road mileage *and* population doubles maintenance costs. Significantly, doubling M alone does not contribute heavily to increasing costs. This is because of the alternate route effect: for a given population more road miles means more alternate routes, only *some* which may be cleared in adverse

Table 7.   Results of *OLS* on Equation (8) in Log Linear Form (1970 Data For 33 States)

Dependent Variable:	Annual State Expenditures on Snow Removal, Salting, Sanding (Millions of Dollars)		
Independent Variables	Dimension	Coefficient	Std Error	*t* level of Significance
Constant	—	$\hat{b}_0 = 0.03$	—	—
Mean annual snowfall	Inches	$\hat{b}_1 = 1.07$	0.13	.995
Road mileage	Thousands	$\hat{b}_2 = 0.22$	0.09	.990
State population	Millions	$\hat{b}_3 = 0.75$	0.06	.995
Number of observations: 33				
F Ratio: 78.64				
$\bar{R}^2 = .89$				

weather. (This, of course, was the reasoning behind our hypothesizing $\beta_2 < 1$.) However, when population increases in the same proportion as road mileage, the new mileage can no longer be termed "alternate" since potential congestion would dictate social need and use of all new mileage. Thus, the homogeneity of TC in M and P has some intuitive appeal.

The highly significant results of this section confirm that highway maintenance costs are indeed climate sensitive. The next section illustrates an application of these results. The application, to an artificial scenario, is meant to demonstrate how research in this field can be useful in actual cost-benefit studies, and also to shed some light on the magnitude of the costs or benefits which may be encountered.

An Application

Let us adopt the following scenario. Transportation officials are understandably interested in investigating the possibilities of widespread use of supersonic transports (SSTs). One policy option is large-scale government involvement in research and production of SSTs. This would lead to the introduction of a large fleet of such vehicles into daily service. It is projected that the first elements of this fleet would start regular service in 1990, and by 2025 a full complement of 1000 SSTs will be in regular operation, transporting passengers and cargo around the globe. The SST is designed to fly in the stratosphere, that layer of atmosphere above the troposphere. The troposphere is where we live and where conventional commercial jetliners fly. While the troposphere cleanses itself of pollutants in a matter of days, pollution of the stratosphere will remain for years. Thus, pollution caused by SSTs will gradually build to high levels in the stratosphere, and this will tend to modify global climate. Scientists predict that the United States will experience a general decline of temperature of 1°C and a general increase in precipitation of 10%. (This scenario, while inspired by a current U.S. DOT project on the climatic impacts of SSTs, bears no necessary relation to that actual investigation. It is merely illustrative.) We shall determine what costs (plus or minus) this climate change will add to U.S. road maintenance expenses.

Our estimate of the impact of the specified climate change on U.S. road maintenance costs is based on the following assumptions:

1. The climatic changes will commence in 1990 and proceed linearly to a new equilibrium (1°C cooler and 10% more precipitation) in 2025.
2. U.S. population will grow at a compound rate of 0.5% per annum until 2025 and then stabilize.
3. Total U.S. road and street mileage will also grow at a compound rate of 0.5% per annum until 2025, and remain stationary thereafter.

4. The estimated model of snow removal costs is directly applicable to all U.S. roads and streets.

We use 1970 data as initial conditions. In 1970, U.S. population was 205 million, U.S. road and street mileage was 3.7 million miles,* and average annual precipitation in the United States was 35 in.† Finally, we need to know what the relation is among snowfall, precipitation, and temperature. The reader will have noted that our highway maintenance model was cast in terms of snowfall, while the climatic modification was described in terms of the more conventional categories of temperature and precipitation. We found that the relation among these variables may reasonably accurately be expressed as

$$S = 199.94 - 3.75\ T + 0.79R \tag{10}$$

where S, T, and R represent snowfall in inches, temperature in degree Fahrenheit, and precipitation in inches, respectively. The estimation of equation (10) is described in the Appendix.

Our objective, in this exercise, is to find the total discounted (to 1974) cost of the climate modification. Because the climate is assumed to remain in its new 2025 equilibrium, the cost stream extends to perpetuity. We shall use an 8% discount rate in the calculation, which proceeds as follows. (The subscript i on any variable refers to calendar year.) The *extra* snow in year i, S_i, is such that

$$S_i = \begin{cases} 0 & i = 1974, \ldots, 1989 \\ \dfrac{9.5}{36} \times (i - 1989) & i = 1990, \ldots, 2025 \\ 9.5 & i = 2026, \ldots, \infty \end{cases}$$

To see this, recall the climate change does not begin until 1990, accounting for the first line. The change being considered is $-1\,^{\circ}C$ and $+ 10\%$ precipitation. Using, from equation (10),

$$\Delta S = -3.75\ \Delta T + 0.79\ \Delta R$$

where $\Delta T = 1.8\,^{\circ}F$ and $\Delta R = 3.5$ in. Recall that equation (10) has T expressed in degrees Fahrenheit, $1\,^{\circ}C = 1.8\,^{\circ}F$, and average annual R for the United States is 35 in.), then ΔS is 9.5 in. This accounts for line three. Finally, line two expresses the assumption that the climate changes linearly over the 36-year period, 1990 to 2025.

---

* See Ref. 6.

† This figure was computed by averaging the mean annual precipitation values for the 196 cities in our sample. Since the sample points were pretty well distributed over the entire United States the figure is not unreasonable. Ideally, of course, one would like a precipitation "function" defined over the coordinates of the United States to work with. Unfortunately, this is not available.

The population in year i is

$$P_i = 205 \times (1.005)^{i-1970} \qquad i = 1970, \ldots, 2025$$
$$= 205 \times (1.005)^{2025-1970} \qquad i = 2026, \ldots, \infty$$

M_i is total U.S. road mileage in year i, in thousands of miles.

$$M_i = 3700 \times (1.005)^{i-1970} \qquad i = 1970, \ldots, 2025$$
$$= 3700 \times (1.005)^{2025-1970} \qquad i = 2026, \ldots, \infty$$

Let Y_i denote the extra snow removal costs in year i. Then, from Table 7,

$$Y_i = 0.03 S_i^{1.07} M_i^{0.22} P_i^{0.75} \qquad i = 1974, \ldots, \infty$$

The present value of this cost stream is given by

$$PV(1974) = \sum_{i=1974}^{\infty} \frac{Y_i}{(1.08)^{i-1974}}$$

The calculation is easily performed by electronic computer. In fact, once the program is written, making small adjustments in the parameters allows a sensitivity analysis to be formed. That is, how sensitive is the result to changes in the projected climate alterations, or to the discount rate, or to population growth? We investigated some of these questions, and the results are presented in Table 8. Note that the estimates are divided into four cases, each corresponding to a different change in temperature. For each case, three different changes in precipitation were considered, thus a total of 12 climate scenarios are represented. For each scenario, costs are discounted to two base years, 1974 and 2000. (This will point up the difference in costs between predicting a change before it occurs and discovering the change after it has begun). Finally, for each base year, three discount rates, 3%, 5%, and 8%, are used in figuring the present value of costs.

The reader will note that the above computation results in a present (1974) value cost of 198 million 1970 dollars.

Several points are made in Table 8. Note that within any case, for either base year, and for any precipitation change, the variation in present value cost among the three discount rates is quite substantial. In fact, in a cost-benefit study, it's conceivable the choice of discount rate could tip the scales one way or the other. Therefore, it is clearly important that attention be given the proper choice of discount rate. Note also that the 2000 discount cost is considerably higher than the corresponding 1974 discount cost. Of course, this is to be expected, since the cost stream begins in 1990. However, it does demonstrate why predicting potential climate changes is superior to discovering them in progress. Finally, note from case 4 that a climate change can result in net social benefits, since the costs are all negative.

Summary, Conclusions, and Recommendations

Climate is a public good. It is both a final consumption good and factor of production. Mankind's sheer numbers coupled with the force of modern technology make inadvertent or designed climate modifications a likely reality for the near future (if it's not already occurring). In fact, the worldwide energy and food problems may make climate control a technological imperative. Upon reflection, it is intuitively clear that climate modifications are not neutral in their effects. Any given change in climate is likely to benefit some production sectors and some consumer groups and is likely to hurt others. The question is inevitably one of trade-offs, and here the economist and the cost-benefit study come to the fore. Whether climate modification should be permitted or even

Table 8. Costs of Highway Maintenance under Selected Parametric Variations of Climate, Discount Rate, and Base Year. (Millions of 1970 Dollars)

Case 1: $\Delta T = -2°C$

		+20%	0	−20%
Precipitation change		+20%	0	−20%
Change in snow (inches)		19.0	13.5	8.0
1974 Present	3%	4209	2877	1587
Value of	5%	1454	994	548
cost stream	8%	415	284	157
2000 present	3%	8635	5904	3256
Value of	5%	4693	3209	1769
cost stream	8%	2538	1735	957

Case 2: $\Delta T = -1°C$

		+10%	0	−10%
Precipitation Change		+10%	0	−10%
Change in Snow (inches)		9.5	6.8	4.0
1974 Present	3%	2005	1374	756
Value of	5%	693	475	261
Cost Stream	8%	198	136	75
2000 Present	3%	4113	2820	1551
Value of	5%	2235	1533	843
Cost Stream	8%	1209	829	456

Table 8   (Continued)

Case 3: $T = -0.5°C$

		+5%	0	-5%
Precipitation Change		+5%	0	-5%
Change in Snow (inches)		4.8	3.4	2.0
1974 Present	3%	980	615	340
Value of	5%	338	214	118
Cost Stream	8%	97	62	34
2000 Present	3%	2010	1263	698
Value of	5%	1092	688	380
Cost Stream	8%	591	371	205

Case 4: $T = +0.5°C$

		-5%	0	+5%
Precipitation Change		-5%	0	+5%
Change in Snow (inches)		-4.8	-3.4	-2.0
1974 Present	3%	-980	-615	-340
Value of	5%	-338	-214	-118
Cost Stream	8%	-97	-62	-34
2000 Present	3%	-2010	-1263	-698
Value of	5%	-1092	-688	-380
Cost Stream	8%	-591	-371	-205

encouraged, to what magnitude, and in what direction are all economic problems. In this chapter, we have proposed an admittedly incomplete methodology for assessing the costs and benefits of some given alteration in climate. We then used the methodology as a guide in investigating the climate sensitivity of the public sector. Two areas were focused on: local public expenditures and state snow removal costs. Statistical analyses allowed us to infer whether, and to what degree, these are climate sensitive. Finally, to illustrate an application of our results, to attempt to gauge the magnitude of costs, and to discover potential problem areas in climate cost-benefit studies, we simulated some climate changes and estimated the impact on U.S. road and street snow removal costs.

In our study of the public sector effects of climate, several conclusions were reached.

1. Of the major categories of *local* public expenditure, education, sewage, and recreation expenditures vary systematically with climate. However, the last

cannot be considered a long run cost of climate modification, while the others can.

2. The model of highway maintenance expenditures is a satisfactory hypothesis. Snowfall, population, and road mileage explain almost all of the variance in state snow removal costs. These costs are homogenous of degree one in average annual snowfall and homogenous of degree one in population and road mileage (jointly).

3. From Table 8, we see that the 1974 present value of snow removal costs beginning in 1990 due to a climate change of $+1\,°C$ and $+10\%$ precipitation ranges above one billion 1970 dollars, depending on the discount rate. Since it is likely that snow removal is only a very small part of the total costs (most of which are presently unidentified), we see that the real cost of climate modification can be significant.

It is customary at this point for the researcher to pay homage to technical points in his area of interest that are yet unresolved, and commend these to future research. As often as not, it seems, it may take some deliberation before he is able to serve up a remaining juicy morsel of a topic. Fortunately, we do not face this problem. In the economics of climate modification, everything remains to be done: methodology, theory, regional studies, industry studies, international studies, private sector, public sector, and so on and on. This may be one of the few areas where economists actually can get a jump on the problem, rather than have it forced down their throats by some future July snowstorm.

Appendix

The Estimation of Equation (10)

This brief section describes the empirical relation among mean annual snowfall, S; mean annual precipitation, R; and mean annual temperature, T. It behooves us to have some forecast, however crude, of snowfall change if we wish to examine the highway maintenance effects of climate modification. The approach we adopt is to regress R and T on S. Our observations are again the average values of those variables for each of the 33 states mentioned above. Collinearity is no problem, since the correlation between R and T is 0.10. The regression equation is $S = \beta_0 + \beta_1 T + \beta_2 R$, as in equation (10). A priori, we expect to find $\beta_1 < 0$ and $\beta_2 > 0$. Table 9 summarizes our results. The coefficients are highly significant and have the expected signs.

Table 9.   Results of *OLS* on Equation (10). [Data Are State Averages of Long Term Site Norms]

Dependent Variable:		Average Annual State Snowfall, in inches		
Independent Variables	Dimension	Coefficient	STD Error	t level of Significance
Constant	—	$\delta_0 = 199.94$	—	—
Average annual state T	°F	$\delta_1 = -3.75$	0.39	.995
Average annual state R	Inches	$\delta_2 = 0.79$	0.14	.995

Number of observations: 33
F Ratio: 54.94
$\bar{R}^2$: .79

[a] The explained variance also seems good, given the admitted crudeness of the model.

References

1. Brazer, *City Expenditures in the United States,* NBER Occasional Paper 66, 1959.

2. Buchanan, *The Public Finances,* Homewood, Illinois: Irwin, 1970.

3. Due and Friedlaender, *Government Finance,* Homewood, Illinois: Irwin, 1973.

4. U.S. Bureau of the Census, *City and County Data Book, 1972,* Washington, D.C.: U.S. Government Printing Office, 1973.

5. U.S. Department of Commerce, *Climatological Data, National Summary,* Annual 1966, Volume 17, No. 13. Washington, D.C.: U.S. Government Printing Office, 1967.

6. U.S. Department of Transportation, *Highway Statistics, 1970,* Washington, D.C.: U.S. Government Printing Office, 1971.

10

Climatic Changes, Local Governments, and Political Problems

RICHARD J. TOBIN
Department of Political Science
State University of New York
at Buffalo
Buffalo, New York

According to Easton,[17] political research seeks to understand the processes through which the authoritative allocation of such values as health, income, and safety are achieved. These authoritative allocations can deprive a person or group of a valued thing already possessed, they can prevent the attainment of values, or they can give some people access to values and deny them to others. Traditionally, political research has focused not so much on the values themselves, but rather on the political and socioeconomic variables that tend to influence the distribution of such values. By their actions or inactions governments establish priorities for the values that are to be available to the public. It is axiomatic that values with low priorities receive lesser attention than those values that are socially defined as important. Crenson,[14] for example, has explored how different political structures and industrial bases can influence the establishment of priorities in regard to air-pollution control.

Besides political and socioeconomic variables, however, other important variables such as climate can influence the distribution of values. This chapter attempts to focus on some possible consequences of climatic modification—especially those consequences of inadvertent man-made changes that influence political activities and the distribution of values in metropolitan areas. First, the essay reviews the major theories of climatic determinism as they pertain to political behavior. Next, it focuses on the nature and distribution of climatic changes in metropolitan areas. Third, the essay attempts to discern some of the political problems and consequences that local governments are likely to face as a result of these climatic changes. The final section is devoted to analyzing future political and socioeconomic developments as they relate to continued climatic changes.

Theories of Climatic Determinism

For thousands of years man has submitted to the idea that his actions are subject to climatic conditions. And, for all these years, people have been describing the relationship between climate and man's activities. In fact, some people have believed that "ideal" or "favorable" climates are directly related to a nation's social, economic, and political progress, or even to a civilization's continued existence. In *Airs Waters Places,* for example, Hippocrates[19] suggested that the best way to ascertain a city's character is to determine which way the city faces. Cities that are sheltered from the north are supposedly exposed to hot winds, which contribute to the residents' proclivity to eczema, dysentery, diarrhea, and chronic fevers. In contrast, Hippocrates believed that cities facing cold winds will have natives who are spare, sinewy, and subject to internal lacerations.

Hundreds of years after Hippocrates, in *The Spirit of the Laws,* Montesquieu[33] again related man's activities to the climate. Montesquieu believed that, since "the tempers of the mind and passions of the heart" are so different in different climates, a country's laws and form of government ought to reflect its climate. Montesquieu argued that monarchical governments are best suited for warm climates, while popular governments are more suited for moderate climates.

Although such theories of climatic determinism as those of Hippocrates and Montesquieu have always had their adherents, it was not until the late 19th and early 20th centuries that the theories reached their peak of popularity with the writings of Mills,[32] Mahaffy,[28] Semple,[40] Koller,[24] Huntington,[22, 23] and Petersen.[36] Mahaffy, for example, believed that hot, tropical climates are predictive of slavery and despotism, and he could cite only a few instances of a country being civilized by its own efforts unless it had either a favorable soil or a favorable climate. Thus, a "happy climate" was the cause of European civilization, or at least Mahaffy believed so.

Despite Mahaffy's bold claims, however, two other scholars gained greater fame and derision for their sweeping theories of climatic determinism. They claimed that climate is almost exclusively responsible for human behavior. Huntington[22, 23] was convinced that some detrimental climatic conditions could lead to a nation's economic and political disintegration while areas of "high climatic efficiency" were likely to assure a country's social and economic progress. According to Huntington, "Many of the great nations of antiquity appear to have risen or fallen in harmony with favorable or unfavorable conditions of climate." In contrast, Los Angeles could never be a center of civilization because of its unvarying mild climate. A hot, muggy climate also condemned the southern United States to backwardness, and Huntington offered ample evidence to support his claim:

One feels it everywhere, for on the trains, at the railroad stations, and at the newsdealers [in the South] it is generally difficult to find the higher grades of magazines. . . . all that I could find was trashy story magazines. . . . Lack of training surely has something to do with the matter, but mental inertia due to lack of climatic stimulus seems to be at least equally important.[42]

Huntington also related variations in the distribution of climatic energy to differences in income, education, human achievement, transportation networks, and size of the population engaged in manufacturing.[22] Huntington concluded his *Civilization and Climate*[22] by reminding his readers that a country's climate is "one of the great reasons why idleness, dishonesty, immorality, stupidity, and weakness of will prevail. If we can conquer climate, the whole world will become stronger and nobler."

Mills also believed that man's social and political activities are dependent on climatic conditions. In *Climate Makes the Man,*[32] Mills asserted that frequent changes in weather drive men "to build skyscrapers, set up great factories, and pursue other energetic activities." Moreover, Mills noted that the growth and development of revolutionary urgings is linked to changes in temperature. On the one hand, warm weather, which supposedly drains a person's energy and vitality, induces people to yield to the power of tyrants. Thus, Mills maintained that a period of extended warmth beginning in 1921 had kept the Italian people submissive to Mussolini's whims and, in fact, excessive warmth, which began in 1929, had initiated a severe and prolonged economic depression. On the other hand, cold climates supposedly have encouraged military aggression and rebellion against despotic repression. And, a gradual cooling in the mid-1930s easily explained Germany's increased aggression that led to World War II.

Man's fascination with the climate has led many scholars to postulate theories of climatic determinism. Many of these theories have excluded virtually all other potential explanatory variables except for climate. As a result, this determinism has been able to "explain" any economic or political occurrence by reference to climatic conditions.

Local Governments and Climatic Changes

Despite the efforts of the climatic determinists, few public officials or social scientists have recognized or appreciated the possible impact of climatic modifications on man's governmental activities. This neglect is apparent when one realizes that public officials and social scientists are only beginning to comprehend the legal, political, and economic significance of such intentional weather modification activities as the seeding of clouds, the diversion of hurricanes, or the suppression of fog and hail. In the last decade, for example, a plethora of new laws, ordinances, and regulations has sought to limit, regulate, or prevent intentional efforts to alter the weather.[41] Today, more than half the American states have laws pertaining to intentional weather modification, and weather modification is a frequent concern of the U.S. Congress, federal agencies, and federally sponsored study commissions as well. In contrast to the growing attention that public officials are now devoting to intentional weather modification activities, only such dramatic, inadvertent climatic changes as the supposed damage to the stratosphere that flights of supersonic transports would cause or the effects of London's episodic killer fogs of the early 1950s have caused public officials to act in the past. Despite the attention given to these dramatic climatic modifications, other gradual and more subtle modifications can have a noticeable impact on governmental functions. Before discussing this impact, however, it is first necessary to describe briefly the nature of climatic changes in metropolitan areas.

The Nature and Distribution of Climatic Modifications

Within Metropolitan Areas. In recent years scientists have increasingly focused attention on man's impact on the climate. Many of these scientists have concluded that man's activities have contributed to significant, long-term changes in the climate, particularly in metropolitan areas. In fact, these changes can be so dramatic and numerous that if Mills were writing today, the title of his book could well be *Man Makes the Climate.* Industrial activity, energy production, alterations in physical features, and the introduction of foreign materials into the urban atmosphere have produced conglomerations that are warmer, wetter, cloudier, foggier, and dirtier than are surrounding rural areas. Tall buildings, paved streets, and parking lots, for example, influence a city's wind flow, energy balance, and precipitation run-off while increased levels of pollution-related aerosols and particulates, which are heavily concentrated in cities, influence visibility and the amount of solar energy that reaches urban areas. Perhaps one of the most noticeable aspects of climatic modification is the difference in urban and rural temperatures. This difference is called the heat-island effect. Changes in the surface, the presence of air pollutants, and the injection of heat into the atmosphere from such artificial sources as transportation, power plants, and industrial processes produce cities that are warmer than their surrounding rural environs. Although the annual mean rural-urban temperature differences may be slight (e.g., 2 to 4°F), Landsberg[25] has found occasional differences as great as 20°F while Bryson and Ross[8] have reported differences as great as 37°F. Changnon,[11] Landsberg,[26] and Peterson[37] describe other climatic changes that occur in cities as a result of man's activities. Their descriptions suggest that man's impact on urban climates is increasing, and a growing concern exists that the present, unintentional modifications are minor compared with the possible macroscale climatic changes of the future. Such projections indicate future changes are likely to have far-reaching societal consequences.

Although it is useful to understand the nature of urban climatic changes, it is equally important to understand the distribution of these changes. By knowing something about the changes' relative distribution throughout a metropolitan area, "it may be possible to understand a good deal about the varying rates and kinds of political behavior that shape public policy."[47] For example, are the effects of a modified climate evenly distributed throughout metropolitan areas? Or, does one group of residents feel the effects more than another? Are some people less likely to bear the consequences of a modified climate merely because of where they live within a city? Because most climatic changes are "absorbed" as diseconomies or negative externalities and operate to the detriment of the affected population, the answers to these questions can have important political consequences in terms of their effect on the distribution of such values as health, income, and safety. On the one hand, if the detrimental effects of a modified climate are evenly distributed throughout a city the changes are non-discriminatory, and no group of urban residents bears a heavier burden than another. On the other hand, if the effects are concentrated in a particular area then that area's residents are disadvantaged compared with residents who live outside the affected area. Both economic and political decisons can play an important role in answering these questions. As an illustration, decisions on the location of intracity highways can benefit commuting suburbanites while penalizing those inner-city residents subjected to "constant noise, higher levels of localized air pollution from exhaust fumes, the glare of lights at night, and increased congestion on some local streets near interchanges."[16]

In order to resolve the question of the relative distribution of climatic effects, it is necessary to determine how a modified climate affects different parts of a city. If a metropolitan area is viewed as a series of concentric circles, with the inner circle representing the central city, and successive rings representing suburbs and surrounding fringe areas, one finds that the most dramatic impact of inadvertent climatic modifications occurs in the central city. Most recent research has found that, in comparison with surrounding rural or suburban areas, central cities tend to have: (*a*) lower wind speeds; (*b*) less sunshine; (*c*) less visibility; (*d*) higher turbidity; (*e*) more air pollution; (*f*) more precipitation; and (*g*) higher temperatures.[8, 25, 37]

Each of these climatic patterns can have a noticeable influence on a central city's residents. Wind speed, for example, can be one of the most important climatic elements. Reduced wind speeds in central cities tend to occur because a built-up city's surface is much rougher than rural or suburban terrain. This roughness creates a frictional drag that affects the wind's structure and causes a major change in the wind's vertical profile so that wind speeds near the surface are reduced by as much as 20 to 30%.[26] Wind speeds can influence humidity, temperature, visibility, and the intensity of pollution. A decrease in wind speed also decreases the dispersion of air pollutants and, in fact, if winds are too light the air can stagnate and produce well-developed dust domes. Similarly, reduced wind speeds decrease a city's ventilation, and this inhibits both the movement of cooler, outside air over a city's center and the evaporative processes within a city.[37]

Differences in temperature (i.e., the heat-island effect) between a central city and its surrounding suburbs are well-documented. Studies of this difference have found that temperatures gradually rise as one moves toward a city's center and that the highest temperatures are found in the inner city, particularly near the highest and densest building and population concentrations.[34] Beside the discomfort associated with higher temperatures, urban heat islands also contribute to the development and maintenance of dust domes. And, the more pronounced the heat island, the higher the wind speed needed to break down the dust dome.[7]

Although suburban areas are not immune to these inadvertent changes, it is apparent that the consequences of these modifications are not evenly distributed within metropolitan areas. Instead, these changes are concentrated in the central or inner cities. Consequently, it is particularly appropriate to examine who is affected by inadvertent climatic modifications.

Who Notices the Changes? Over the last three or four decades a tremendous change has occurred in the composition of metropolitan areas in the United States. Two simultaneous migrations have characterized this change. First, between 1940 and 1970 over 3,500,000 Blacks migrated from the South primarily to the largest central cities of the North. More recently, of all Blacks who changed residences between 1970 and 1973, over 60% settled within the central cities of the Nation's largest metropolitan areas. This movement to the central cities has also included disproportionate numbers of the poor, Cubans, Puerto Ricans, and Mexican-Americans. Once these Blacks and other minority-group members do settle in the central cities, they are likely to settle near the center of the city, that is, in those areas most susceptible to the consequences of inadvertent climatic changes. In contrast to the Blacks' migratory patterns, only one-fourth of all white Americans who changed residences between 1970 and 1973 settled in the largest central cities, and many of them settled in the more expensive single-family dwellings or upper-income apartment buildings near the central cities' periphery. As a result of these changes, nearly 60% of all black Americans now live in the central cities of metropolitan areas, and this percentage is likely to increase substantially in the next generation.

Second, the last several decades have also seen a massive out-migration of well-to-do whites from the central cities to the suburbs. Between 1960 and 1970, 40 of the Nation's 72 largest central cities experienced an actual decline in their white populations, while all but six of these same cities' suburbs experienced a growth in white populations.[1] These population shifts are particularly acute in the Northeast, which is heavily urbanized, as well as in such midwestern cities as Detroit, Cleveland, and St. Louis. Because of this out-migration, only one of four white Americans now lives in a central city.

In sum, the detrimental effects of modified climates are unevenly distributed within metropolitan areas and among metropolitan populations. This conclusion is especially important when one realizes that the allocation of such diseconomies is largely within the control of public officials who can regulate or modify many of the causes of inadvertent climatic changes. Indeed, the presence of diseconomies creates a presumption toward some form of governmental activity.[38] Coincidentally, the absence of public regulation can also affect the distribution of diseconomies among local governments, as noted below.

Outside Metropolitan Areas. If the residents of central cities are the most likely to suffer the fates of a modified climate, another group of residents outside metropolitan areas also suffers because of man's influence on the climate. This second group of residents, however, is often physically removed from the metropolitan area that causes the changes.

Man's activities in metropolises such as Cleveland, Chicago, St. Louis, and New York apparently influence climatic patterns in neighboring areas. Changnon,[12] for example, has found that at La Porte, Indiana, which is approximately 30 miles downwind from Chicago, the amount of precipitation and the number of days with hail and severe thunderstorms have increased markedly in the last 50 years. La Porte has nearly 40% more thunderstorms and nearly 250% more hailstorms than the neighboring countryside. Changnon also found a close correspondence between the variations in La Porte's precipitation and the number of smoke-haze days in Chicago. Other surveys by Wyckoff,[50] Huff and Changnon,[21] and Bryson and Ross[8] have reported similar downwind effects. In each of the studies cited the researchers have suggested that industrial and combustion processes contribute to the modified climates of the downwind areas. In these cases, those who are responsible for the changes are not the same ones who bear the costs.

Downwind modifications are particularly disconcerting to those affected because they usually have no effective recourse should they attempt to prevent the climatic changes. In La Porte's case, both physical distance and political boundaries separate the residents from the probable causes of the climatic changes. Chicago's and Gary-Hammond's metropolitan areas comprise parts of 2 states, 8 counties, 136 townships, 284 municipalities, and over 800 other governmental units. Such fragmentation characterizes most metropolitan areas and contributes to an absence of political responsibility. Under these circumstances, the actions of industrial and commercial concerns in Chicago are free from regulatory control by La Porte's public officials. Indeed, although federal intervention is possible, it is highly unlikely that any form of federal political or regulatory action would be sufficient to eliminate Chicago's downwind climatic effects on La Porte. In short, the likelihood of effecting a substantially different pattern of climatic externalities in La Porte is remote.[47]

Political Problems and Consequences

If, as it is widely assumed, changes in weather and climate affect man's economic, agricultural, and recreational activities, then it is reasonable to assume that such meteorological changes also affect man's political and governmental activities. Unfortunately, however, few researchers have explicitly explored the relation between climatic changes and the authoritative allocation of values.

Political demands are one element that can influence this allocation. The manner in which governments respond to such demands substantially determines how important values will be distributed. Generally, when public officials do respond to demands, specific decisions result in the allocation of values. For example, when public officials accept citizens' demands for a high-cost pollution-abatement program, this acceptance can influence the concerned citizens' sense of political efficacy, their degree of satisfaction with the political system, and the distribution of health and financial resources. In contrast, the same decision may negatively affect previously unencumbered industries, which may be required to install expensive and nonproductive pollution-control devices.

Few governments, however, have the capacity to satisfy or respond to all demands. Thus, some demands do not become issues subject to resolution by intentional decisions because public officials do not recognize or appreciate the importance of the demand because of a lack of time, resources, or information. In these instances choices result from nondecisions, and the choices simply "happen."[5] This pattern seems to characterize the posture of most local governments toward inadvertent climatic changes.

Climatic Changes and Political Activity

In the realm of climatic changes and political activity, many unanswered questions abound. As Haas[18] accurately asserts, man has developed a comprehensive set of behaviors that are intended to cope with a climate that he has experienced over a long period. If a city's climate is significantly altered, then some kind of readjustment in traditional behavior patterns is likely to occur. But, what is the nature of this readjustment as it relates to political activity?

Writing in the mid-1920s, Barnhart[6] attempted to answer this question when he related Nebraska's drought in 1890 to the rise in the number of independent voters and to the creation of the Populist Party. Since Barnhart's essay few political scientists have related changes in either weather or climate to political behavior, that is, behavior that affects or is intended to affect a government's decisional outcomes.[31] Haas,[18] Mann,[29] and Holden[20] all speculate that meteorological elements affect such acts of political participation as voter registration and electoral turnout, but none of the authors specifies the relationship among the variables. Milbrath,[31] in contrast, has remarked that a person who has to "go out in inclement weather has a greater energy and time cost for his participation than does a person without such barriers." Thus increases in the frequency of "unfavorable" weather are likely to be associated with declining levels of political participation. Because of the great number of variables that can affect voting and registration, however, it may be especially difficult to identify the particular influence of climatic changes on political activity.

If political behavior is intended to influence a government's decisions, then riots certainly qualify as an expression of public behavior intended to bring about change. Although many variables can influence the incidence of riots, long periods of hot, humid, uncomfortable weather can produce anxiety and psychological stress.[4, 30] Two of the by-products of stress are fatigue and disturbance of behavior, and

. . . with fatigue there tends to be a loss of perspective where minor happenings become intensified, or when they lose their true significance. For example, in the United States many race riots are said to start from some trivial incidents between a few Negroes and the police. The situation often expands rapidly and a mob may form. Then, through the application of some form of stress, tension is built up to the point where it explodes into a riot, and it is suggested that the development of this stress can be initiated or at least heightened by changes in various weather parameters such as temperature, rainfall, humidity and possibly atmospheric pressure.

In the United States such a tension often arises out of the squalor and poverty of Negro ghettoes, and here weather could be said to act as a spark to light the already-present fuel, but it is not the fuel itself.[30]

Maunder[30] suggests that such a chain of events led to Detroit's riots in June of 1943, and to East Harlem's riots in July of 1967. Other researchers also give brief attention to the "long hot summer mentality." Although empirical evidence is lacking, Cobb and Elder suggest that:

Around June, decision-makers collectively brace themselves for a series of violent outbursts in the ghettoes and provide token concessions to prevent an insurrection or an accelerated level of violence. As the summer ends, minority group demands lose their high priority and indeed may disappear from the docket until the following June.[13]

Climatic changes can also affect political activity by increasing the number of citizens who focus attention on the results of such changes. At present, relatively few urban residents realize that, by "agreeing" to live in a city, they have also "agreed" to live in a modified climate. If inadvertent climatic changes should intensify and if metropolitan residents become more aware of the gradual climatic changes to which they are subjected, then such changes will likely lead to numerous demands for public action. Those people who must remain or who choose to remain in a metropolitan area may decide that they will no longer tolerate massive changes in temperature, precipitation, visibility, and so forth. In the past, public officials have generally been free from such demands and have not had to consciously allocate the consequences of a modified climate, some of which are described in the sections that follow.

Climate, Crime, and Police Behavior

Sociologists, criminologists, and political scientists have long sought to explain the causes of crime. One approach has attempted to explain crime in terms of changes in weather or climate. Advocates of the "thermal theory," for example, claim that hotter temperatures are likely to induce crimes against persons while colder temperatures are likely to induce crimes against property. A study of crime in Germany tends to support at least one part of this theory. Aschaffenburg[3] found that assaults, obscene acts, and sexual crimes all reached their yearly peaks in the warmest months. Dexter[15] also examined the incidence of crime in relation to climate and concluded that crimes of violence were most frequent on clear days, when winds were mild, and when humidity and atmospheric pressure were low. In contrast, Dexter also found that rain, cloudiness, high pressure, and high humidity were associated with fewer crimes of violence. Needless to say, the causal relationship between crime and climate hypothesized by the thermal theory has been the source of significant controversy.

Although strict adherence to the thermal theory may be questionable, some recent evidence does suggest a relation between climatic conditions and the incidence of certain crimes. In its *Uniform Crime Reports,* the Federal Bureau of Investigation[44] lists "climate, including seasonal weather conditions," as one variable that affects the type and volume of crime. In addition, the *Reports* include charts that reflect monthly variations in the incidence of crime. These charts show that murders, forcible rapes, and aggravated assaults all reach their peak in the late summer, when temperatures are the highest.

Perhaps the most thorough studies of the relation between crime and climate are those by Will and Sells,[39, 49] who have examined the relationship between 11 measures of weather and 57 categories of telephone calls (e.g., calls to report assaults, accidents, burglaries, robberies, etc.) handled by the police department of Fort Worth, Texas. Through the use of factor analysis the authors found that almost one-fourth of the police variables were significantly correlated (i.e., $p < .01$) during at least one period of the day with a factor composed of high rain and low visibility. More important, nearly half of the police variables were significantly correlated with a weather factor representing high temperature and low atmospheric pressure. For certain periods of the day, Will and Sells found that as the temperature/pressure factor changed, sizeable increases in the number of prowlers, disturbances, and investigations occurred, but that the number of minor accidents and stolen cars decreased. As a result of their findings, the authors concluded that a general increase in police activities occurs as temperature increases and as pressure decreases.

Will and Sells' findings are noteworthy in regard to man's inadvertent climatic modifications. The two weather factors (i.e., temperature/pressure and rain/visibility) that the authors identified as being closely related to changes in police calls are the same climatic elements most likely to be modified by man's activities. Thus, these data suggest that a city's modified climate is likely to contribute to an increase in police calls. Similarly, if changes in temperature/pressure are correlated with the number of police calls, then it is also likely that those areas most affected by climatic changes (i.e., the center of metropolitan areas) will produce the greatest increase in police calls. Although many variables contribute to requests for police action, the evidence suggests that climatic changes exacerbate previously existing conditions, especially when urban-rural temperature differences are maximized. Further research is needed to determine the nature of the relationship between crime, climatic conditions, and requests for police action. Such research could provide useful insights for metropolitan police departments. For example, simulation models could be constructed on the basis of anticipated temperature/pressure changes and used to assist police departments and municipal governments in determining the most effective distribution of their resources.

Climate, Health, and Property

Direct Effects.   When man's activities change a city's climate, the comfort of the inhabitants is also changed. Although many people enjoy warm weather, the heat-island effect can raise temperatures beyond desirable levels, especially during humid summer evenings and nights. More important than comfort, however, is that increases in temperature due to the heat-island effect can also lead to a sizeable change in the number of deaths and heat stroke victims. Buechley, Van Bruggen, and Truppi,[9] for example, examined a heat wave that occurred in New York City's metropolitan region in the summer of 1966. They found that temperatures above 90°F were related to excess mortality due to heat. At 95°F the excess mortality was 27% higher than the yearly average; at 100°F the excess mortality was 75% higher. Because of the differences in temperatures between the city's core and suburbs, Buechley and his co-workers estimated that approximately 150 to 200 extra deaths had occurred in the city the day following the highest temperatures. In brief, the authors found that urban heat islands are coincident with a region of high mortality.

Indirect Effects.   Two factors that determine a pollutant's concentration within an urban area are the magnitude of the local emission source for that pollutant and the prevailing meteorological conditions. Pollutants from fuel combustion, automobile exhaust, and industrial processes are heavily concentrated in metropolitan areas and, once the pollutants are in the atmosphere, their dispersion is dependent on such meteorological variables as temperature, wind speed, and atmospheric stability—all influenced by man's activities. As an illustration, Padmanabhamurty and Hirt[35] found that increases in the intensity of heat islands are related to increases in haze and concentrations of sulfur dioxide. A city's excessive heat can also contribute to temperature inversions, which shut off ventilation systems. During inversions, pollutants tend to accumulate in the stagnant air of the inner city. As a consequence dispersion is minimized, and the pollutants' impact on health and property is focused on a relatively small part of the metropolis, that is, that part of the city increasingly inhabited by the poor, the Black, and other minority-group members. Again, the consequences of a modified climate are unevenly distributed. Moreover, those people who are disadvantaged by increased levels of climate-related pollution are largely the victims of pollution produced by others.

Health.   Although social and economic differences can contribute to variations in the incidence of disease among residents of inner cities and suburbs, much evidence indicates that air pollution is an important contributory cause to illness and even death. Many studies have documented a close relationship between levels of air pollution and the increased incidence of heart and chronic respiratory disease and stomach and prostate cancer. Other studies have linked air pollution with total death rates. At the extreme, most people are familiar with the inversion-related deaths in Belgium's Muese Valley (1930), Donora, Pennsylvania (1948), and London, England (1952). In London, the number of deaths above normal during the inversion was 2700, and the number of deaths remained considerably above average for several weeks thereafter.

The increased incidence of death and disease due to the heat-island effect and climate-related pollutants can have several important consequences for local governments. First, unhealthy people are less likely than healthy people to make a contribution to a city's prosperity. In comparison with the healthy, the ill are less likely to work and less likely to pay equivalent taxes. Lave and Seskin[27] estimate that air pollution is responsible for nearly half the lost income and current medical expenses associated with bronchitic morbidity and mortality. Second, because inner-city residents tend to be less wealthy than their suburban neighbors, sick inner-city residents are more likely to rely on municipal clinical and medical facilities, which are frequently supported by tax revenues instead of by patient fees. In sum, these circumstances serve to exacerbate a situation in which more people demand more municipal services while providing fewer resources.

Property.   While it is well-documented that air pollution damages property, less evidence is available about the relation between climate, pollution, and property damage. Nevertheless, several studies reveal that climatic variables tend to increase the amount of property damage due to pollution. Because climatic variables can prevent the dispersion of pollutants, property damage within a central city is greater than would be the case if pollutants could disperse normally. Moreover, although precipitation scavenging is one of the major means of removing pollutants from the air, the emission of sulfuric and nitrogenous compounds into an urban atmosphere dramatically increases the acidity of precipitation. Recent estimates for the eastern United States suggest that the acidity of rainfall has increased 100 to 1000 times the normal levels in the last 20 years. Acidic rainfall can damage freshly applied paint, can make it dry more slowly, can make it more permeable to water, and can otherwise accelerate corrosive damage to metals, concrete, and building stones. Waddell[48] has concluded that levels of air pollution are inversely related to median-property values. For all housing units within a metropolitan area, Waddell has estimated that air pollution reduces median-property values in the United States by $350, for a total damage of nearly 6 billion dollars. Although the total damage to property due to air pollution cannot be solely attributed to climatic causes, it is likely that a modified climate contributes to a sizeable portion of the damage.

A decrease in property values has a noticeable impact on local governments.

The property tax, which is based on assessed-property values, is the single most important source of revenue for local governments in the United States. In 1972–1973, the property tax accounted for 12 billion dollars or over one-fourth of local governments' total revenues. Consequently, a decrease in property values can seriously affect the ability of a local government to finance demands for new services such as health care or pollution control. Additional increases in property taxes to offset the costs of the pollution-related damage tend to encourage more citizens to move to the suburbs while leaving behind those people most in need of services and least able to pay for them.

Other Potential Consequences

There are several other possible consequences that result from inadvertent climatic changes. In most instances, these consequences have not been the subject of extensive investigation and, therefore, their probable occurrence is usually ignored by public officials. Some of these consequences include the following:

1. One possible change is an increase in the pollution of groundwater, which serves as a major source of municipal-water supplies. Changnon[12] reports that St Louis' rapid increase in groundwater pollution appears to be related to greater atmospheric pollution and greater amounts of rainfall. In some regions, such as the southwestern United States, a shortage of clean water can actually retard continued economic growth.

2. Another water-related problem concerns the likelihood of floods. Because the process of urbanization tends to intensify the number of heavy rainstorms, the frequency and magnitude of flood-producing showers is likely to be far greater than that experienced in neighboring fringe areas. A greater incidence of heavy rainstorms also affects the design and placement of urban sewage systems, which are frequently tied in with drainage systems.

3. If a modified climate makes a city less desirable to live in, then such a climate is likely to encourage an additional exodus from a city while discouraging the in-migration of those people who have a choice where they live (i.e., primarily well-to-do whites who can choose between central-city and suburban neighborhoods). Obviously, both migrations act to decrease the tax revenues available to a city's government.

The Prospects for Change

If man's activities in metropolitan areas are responsible for changes in the climate, then what are the prospects for eliminating these unwanted modifications? First, it is clear that the Nation's future population growth will be concentrated in metropolitan regions. A report by the Advisory Commission on Intergovernmental Relations[2] projects that 27 urbanized areas in the United States will increase their populations by over 65 million by the end of this century. By the year 2000, four major supermegalopolises comprising only 8% of the land will hold nearly 60% of the country's population. In short, the future holds promise for substantially more metropolitan growth, which will be concentrated on small land areas. As a result, today's suburbs will increasingly share the central city's climate.

Continuing metropolitan growth is important because it is directly related to incremental climatic changes—the greater the growth of cities, the greater the intensity of the urban heat island, the magnitudes of floods and pollution, the number of fog and hail days, and the frequency of thunderstorms. In contrast, metropolitan growth is inversely related to ventilation and wind velocity.

Second, it is clear that urban transportation systems and the production of electricity, which are two of the leading causes of climatic change, will be more important contributors to a modified climate than they are now. Conventional fossil-fueled power plants, for example, currently generate more waste heat than useful electrical energy and also evaporate as much as 14,000 gal of water per day. Moreover, consumption of electricity in the United States increases at a rate of approximately 4 to 7% per year, and the annual per capita consumption of electricity is expected to rise from approximately 9850 kwh in 1975 to over 32,000 kwh in the year 2000. Although fossil-fueled plants' efficiency may gradually improve, the increased use of electrical energy is likely to result in large increases in pollution, evaporation, and waste heat.

As the United States gradually turns to nuclear power, however, it will find that such power plants are even less efficient and eject more heat than conventional plants. In early 1976 there were only 58 nuclear plants in operation in the United States, but by 1985 the Nation is expected to have between 270 and 340 nuclear-power plants in operation. After 1985 nuclear plants will represent approximately 90% of all additions to generating capacity.[43]

Because of limitations on pressure and temperature that are not present in conventional plants, nuclear facilities operate at lower efficiencies than do fossil-fueled plants. In other words, for an equivalent amount of electrical energy delivered, nuclear plants release more waste heat to the environment than do fossil-fueled plants. Even though nuclear plants' thermal efficiencies may improve, it is not certain that they will attain the same efficiencies that conventional plants now possess. Similarly, even though cooling towers for nuclear plants can offer a partial solution for excess waste heat, these towers can evaporate as much as 40% more water than fossil-fueled plants. Thus, both the excess heat and water associated with nuclear plants can be expected to increase and to dramatically change the urban climate.

Metropolitan transportation systems also contribute to climatic changes because automobile emissions are a major source of air pollution and a leading source of waste heat. Although state and federal governments have set pollution-abatement standards, automobile manufacturers have been highly successful in delaying the standards' implementation, and total compliance will neither eliminate all pollutants nor reduce waste-heat discharges. Moreover, Americans' fondness for the automobile has been expressed at the expense of other forms of transportation:

In the past 25 years [1949 to 1973], the number of passengers carried by the public transit modes has declined from 23 billion passengers annually to 6.8 billion. At the same time, the use of the automobile in urban areas has burgeoned. In the decade of the 1960s alone, while transit ridership was dropping from 9.4 to 7.3 billion passengers annually, automobile passenger miles in urban areas increased from 423 billion to 737 billion.

According to the Department of Transportation's 1972 National Transportation Report, almost 94 percent of all urban passenger miles of travel today is by the private automobile, with buses accounting for 2.7 percent, subways and other rail systems 2.2 percent, commuter rail and taxicabs less than 1 percent each.

Passenger mile statistics somewhat underestimate the relative importance of taxicabs. In 1970, taxicabs actually hauled more passengers, albeit for shorter distances, than the combined total of the subway/surface rail, trolley coach and commuter train modes, and more than one-half as many as the bus mode. In many communities, as bus service has disappeared altogether, taxicabs are the only form of public transportation available.[46]

The number of households that owned two or more cars more than doubled between 1960 and 1973, and studies show that as vehicle ownership increases, the mileage each vehicle is driven also increases. For example, a one-car family drives their vehicle an average of 10,800 mi annually, but for a two-car family, the average is 12,000 mi *per vehicle*.[46]

Governmental projections suggest that these trends will persist. A recent study by the United States Department of Transportation[45] predicted that the number of motor-vehicle registrations will increase by nearly 50% between 1970 and 1990, and that the number of vehicle miles traveled in urban areas will increase more than twofold in the same period. The same study also suggested the need for nearly 18,000 mi of additional expressways in metropolitan areas by 1990, compared with about 8000 mi in 1968. Should these projections be fulfilled, then it is probable that automobile traffic will claim an increased share of the responsibility for climatic changes.

Third, the American value system contributes to further climatic changes by encouraging economic and industrial expansion. Subsidies, price supports, investment-tax credits, and accelerated-depreciation allowances are all intended to spur further industrial growth and resource development. Understandably, in the face of growing demands for public services, local governments are also interested in the larger tax bases and increased governmental revenue that industrial expansion can generate. Given a choice between economic development and the reduction of subtle climatic changes, public officials are unlikely to choose the latter. It is hoped that the choice between alternatives will not be so dramatic, because there may be some policies that will accomodate both economic growth and a reduction of climatic changes.

Finally, despite the many problems that a modified climate can cause for an urban government, the near future is unlikely to bring any major changes in the action or attitudes of public officials. These officials must first be convinced that consideration of inadvertent climatic changes is something with which they should be concerned. Political decision makers have seemingly excluded this topic from their agendas and are likely to continue to do so. Governmental dockets at all levels tend to be crowded with prior problems, and it is especially difficult for such new issues as inadvertent climatic changes to be placed on an agenda. According to Cobb and Elder,[13] new issues do not reach governmental agendas unless there is: "(a) widespread attention or at least awareness; (b) shared concern of a sizeable portion of the public that some type of action is required; and (c) a shared perception that the matter is an appropriate concern of some governmental unit and falls within the bounds of its authority."

Scientists concerned with controlling climatic changes have made little progress with any of these requirements. For example, although scientists are collecting an impressive array of evidence linking urbanization and climatic changes, *almost none of this information is effectively transmitted to the people who must cope with such changes*—elected public officials. As Burton[10] notes, "It is not simply a matter of reporting such information in scientific terms but of translating and interpreting understandings in such a way that they may be incorporated into a decision process." Unless this transmission and incorporation occurs more effectively, inadvertent climatic changes will continue to be the subject of few public official's decisions.

References

1. Advisory Commission on Intergovernmental Relations, *City Financial Emergencies: The Intergovernmental Dimension,* Washington, D.C.: U.S. Government Printing Office, 1973.

2. Advisory Commission on Intergovernmental Relations, *Urban and Rural America: Policies for Future Growth,* Washington, D.C.: U.S. Government Printing Office, 1968.

3. Aschaffenburg, G., *Crime and Its Repression,* Translated by A. Albrecht. Boston: Little, Brown, 1913.

4. Bach, W., "Urban Climate, Air Pollution, and Planning," in T. R. Detwyler and M. G. Marcus, Eds., *Urbanization and Environment: The Physical Geography of the City,* Belmont, Calif.: Duxbury, 1972.

5. Bachrach, P. and Baratz, M. S., *Power and Poverty: Theory and Practice,* New York: Oxford University Press, 1970.

6. Barnhart, J. D., "Rainfall and the Populist Party in Nebraska," *American Political Science Review,* Vol. 19, (August 1925), pp. 527–540.

7. Bryson, R. A., "Climatic Modification by Air Pollution," in N. Polunin, Ed., *The Environmental Future,* New York: Barnes and Noble, 1972.

8. Bryson, R. A., and Ross, J. E., "The Climate of the City," in T. R. Detwyler and M. G. Marcus, Eds., *Urbanization and Environment: The Physical Geography of the City,* Belmont, Calif.: Duxbury, 1972.

9. Buechley, R. W., VanBruggen, J., and Truppi, L. E., "Heat Island = Death Island?" *Environmental Research,* Vol. 5, (March 1972), pp. 85–92.

10. Burton, I., "Issues in the Design of Social Research for the Management of Atmospheric Resources," in W. R. D. Sewell, et al., *Modifying the Weather: A Social Assessment,* Victoria, British Columbia: University of Victoria, 1973.

11. Changnon, S. A., Jr., "Atmospheric Alterations from Man-Made Biospheric Changes," in W. R. D. Sewell, et al., *Modifying the Weather: A Social Assessment,* Victoria, British Columbia: University of Victoria, 1973.

12. Changnon, S. A., Jr., "Urban-Industrial Effects on Clouds and Precipitation," *Preprints of the Fourth National Conference on Weather Modification, Logan, Utah, August 13–31, 1973,* Boston: American Meteorological Society, 1974.

13. Cobb, R. W. and Elder, C. D., *Participation in American Politics: The Dynamics of Agenda Building,* Boston: Allyn and Bacon, Inc., 1972.

14. Crenson, M. A., *The Un-Politics of Air Pollution: A Study of Non-Decisionmaking in the Cities,* Baltimore: Johns Hopkins Press, 1971.

15. Dexter, E. G., *Weather Influences,* New York: Macmillan, 1904.

16. Downs, A., *Urban Problems and Prospects,* Chicago: Markham, 1970.

17. Easton, D., *A Framework for Political Analysis,* Englewood Cliffs, N.J.: Prentice-Hall, 1965.

18. Haas, J. E., "Sociological Aspects of Human Dimensions of the Atmosphere," in National Science Foundation, *Human Dimensions of the Atmosphere,* Washington, D.C.: U.S. Government Printing Office, 1968.

19. Hippocrates, *Airs Waters Places,* in *Loeb Classical Library,* Translated by W. H. S. Jones, London: Heinemann, 1962.

20. Holden, M., Jr., "Politics and Weather Modification," in W. R. D. Sewell, et al., *Modifying the Weather: A Social Assessment,* Victoria, British Columbia: University of Victoria, 1973.

21. Huff, F. A. and Changnon, S. A., Jr., "Climatological Assessment of Urban Effects on Precipitation at St. Louis," *Journal of Applied Meteorology,* Vol. 11, (August 1972), pp. 823–842.

22. Huntington, E., *Civilization and Climate,* 2nd ed., New Haven: Yale University Press, 1922.

23. Huntington, E., *The Pulse of Progress,* New York: C. Scribner's Sons, 1926.

24. Koller, A. H., *The Theory of Environment,* Menosha, Wisc.: Collegiate Press, 1918.

25. Landsberg, H. E., "The Climate of Towns," in W. L. Thomas, Ed., *Man's Role in Changing the Face of the Earth,* Chicago: University of Chicago Press, 1956.

26. Landsberg, H. E., "Climates and Urban Planning," in *Urban Climates,* Geneva: World Meteorological Organization, 1970.

27. Lave, L. B. and Seskin, E. P., "Air Pollution and Human Health," *Science,* Vol. 169, (August 21, 1970), pp. 723–733.

28. Mahaffy, J. P., *Twelve Lectures on Primitive Civilizations,* London: Longmans, Green and Co., 1869.

29. Mann, D. E., "Human Dimensions of the Atmosphere from the Perspective of a Political Scientist," in National Science Foundation, *Human Dimensions of the Atmosphere,* Washington, D.C.: U.S. Government Printing Office, 1968.

30. Maunder, W. J., *The Value of Weather,* London: Meuthen, 1970.

31. Milbrath, L. W., *Political Participation,* Chicago: Rand McNally, 1965.

32. Mills, C. A., *Climate Makes the Man,* New York: Harper and Bros., 1942.

33. Montesquieu, C., *The Spirit of the Laws,* in *Great Books of the Western World,* Vol. 38, Chicago: Encyclopedia Britannica, 1952.

34. National Science Board, *Patterns and Perspectives in Environmental Science,* Washington, D.C.: U.S. Government Printing Office, 1972.

35. Padmanabhamurty, B. and Hirt, M. S., "The Toronto Heat Island and Pollution Distribution," *Water, Air, and Soil Pollution,* Vol. 3, (1974), pp. 81–89.

36. Petersen, W. F., *Lincoln-Douglas: The Weather as Destiny,* Springfield, Ill.: Thomas, 1943.

37. Peterson, J. T., *The Climate of Cities: A Survey of Recent Literature,* Durham, N. Car.: National Air Pollution Control Administration, 1969.

38. Samuelson, P. A., *Economics,* New York: McGraw Hill, 9th ed., 1973.

39. Sells, S. B. and Will, D. P., Jr., *Accidents, Police Incidents, and Weather: A Further Study of the City of Fort Worth, Texas, 1968,* Fort Worth: Institute of Behavioral Research, Texas Christian University, 1971.

40. Semple, E. C., *Influences of Geographic Environment,* New York: Henry Holt, 1911.

41. Sewell, W. R. D., et al., *Modifying the Weather: A Social Assessment,* Victoria, British Columbia: University of Victoria, 1973.

42. Silverberg, R., *The Challenge of Climate,* New York: Meredith, 1969.

43. U.S. Atomic Energy Commission, *Nuclear Power 1973–2000,* Washington, D.C.: U.S. Government Printing Office, 1972.

44. U.S. Department of Justice, Federal Bureau of Investigation, *Uniform Crime Reports in the United States,* Washington, D.C.: U.S. Government Printing Office, 1973.

45. U.S. Department of Transportation, *1972 National Highway Needs Report, Part II,* Washington, D.C.: U.S. Government Printing Office, 1972.

46. U.S. House of Representatives, Committee on Public Works, *Urban Transportation (Dilemmas at a Time of Decision)*, 93rd Cong., 1st Sess., 1973.

47. Wade, L. L., *The Elements of Public Policy*, Columbus, Ohio: Charles E. Merrill, 1972.

48. Waddell, T. E., *The Economic Damages of Air Pollution*, Washington, D.C.: U.S. Government Printing Office, 1974.

49. Will, D. P., Jr. and Sells, S. G., *Prediction of Police Incidents and Accidents by Meteorological Variables*, Forth Worth, Tex.: Institute of Behavioral Research, Texas Christian University, 1969.

50. Wyckoff, P. H., "NSF Weather Modification Program Highlights for FY-69, 70 and 71," in U.S. Department of Commerce, National Oceanic and Atmospheric Administration, *Summary Report: Weather Modification, Fiscal Years 1969, 1970, 1971*, Washington, D.C.: U.S. Government Printing Office, 1973.

11

The Capital Costs of
Climatically Induced Shifts
in Agricultural Production:
the Example of the
American Corn Belt

JOHN H. NIEDERCORN
Department of Economics
University of Southern California
Los Angeles, California

Long-term climatic change induced by aircraft of various types passing through the stratosphere is likely to affect the earth and its inhabitants in many ways. Potentially one of the most significant consequences to be expected is a permanent drop in temperature and a massive set of concomitant worldwide shifts in patterns of agricultural land use.

Background

Recent research has shown that the future availability of food, fiber, and wood resources in certain areas is highly sensitive to small variations in temperature and rainfall. A relatively small lowering of mean temperatures

259

would probably shift the Corn Belt significantly to the south and drastically reduce the production of wheat in Canada and the Soviet Union, and the grain output of northern China. Production of these commodities would necessarily shift southward. It would tend, to some extent, to displace the warm-temperature crops such as cotton, which, in turn, would shift further to the south. However, the whole network of highways and central places (agriculturally oriented urban centers) that has grown up in the most northerly regions would then very likely be partially abandoned. In more southerly regions, existing central places would necessarily be expanded and new ones developed. This tendency toward increased labor force, capital stock, and population in southern regions would likely be accentuated by the continuing long-term increase in world demand for foodstuffs, and the higher world market prices that now appear to be inevitable for agricultural commodities.

Capital Stock Shifts

This study attempts to measure the costs of shifts in the network of highways and central places necessitated by the changes in the location of agricultural production. Simply stated, it involves the estimation of capital requirements for the construction of new central places and the expansion of those that already exist. These capital requirements must be understood in a net sense, that is, new capital construction generated by climatic and resultant land-use changes minus the construction that would have taken place without the changes.

The Corn Belt

Originally, the author intended to estimate the capital costs associated with changes in the location of production for a number of crops including corn and wheat. Financial and time constraints limited the study to these two most important crops. However, since different varieties of wheat exist, each best suited to differing climatic conditions, it is difficult to specify how wheat production would shift if the most northerly farms had to be abandoned. At any rate, it is clear that the production of wheat would not be concentrated in a compact, well-defined geographical region. For these reasons, the attempt to analyze capital costs resulting from shifts in wheat production was abandoned.

Fortunately, corn presents none of the difficulties associated with wheat. In addition, the Corn Belt is clearly the most important agricultural region in the United States. Therefore, the author decided to concentrate this study on estimating the capital costs necessitated by a shift in the Corn Belt. A list of the states and counties included in the Corn Belt with and without climatic change is given in Appendix A.

Even though the Corn Belt produces only a part of total U.S. agricultural output, and the United States accounts for only a fraction of the world's total, it

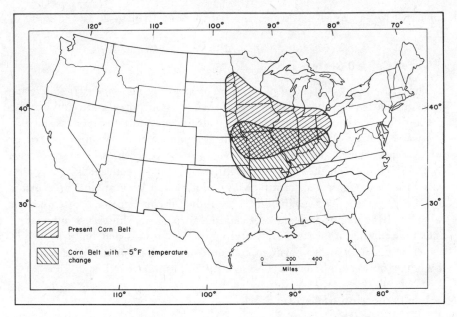

Figure 1. Present and future Corn Belt area with 5°F (2⅘°C) July mean temperature decrease (1).

can be argued that the capital costs incurred through a shift in the Corn Belt may be viewed as a good indicator of whether or not a potential worldwide problem exists. If these costs appear small, it might be reasonably safe to conclude that the likely worldwide costs would not be disastrous. On the other hand, if costs associated with a shift in the Corn Belt appear to be large, a much more detailed study of capital costs on a worldwide basis would seem to be in order.

Methodology

Overview

The basic methodology of this study is straightforward. Capital stocks in the Corn Belt that would come into existence with climatic change are projected to the years 2000 and 2025 under the two assumptions of continuation of the existing climate and climatic change. The differences between the annual net investments under the two assumptions are then calculated. These differences for the years 2000 to 2025 are discounted back to the base years of 1974 and 2000. The resultant estimates represent the total costs of climatic change as perceived in the years 1974 and 2000, respectively. These costs are then converted to an annual basis over the two periods 1974 to 2025 and 2000 to 2025.

Three Capital Sectors

The creation of new central places involves not only the construction of grain elevators, harbors, railway yards, roads, highways, airports, and retail outlets, but also the provision of housing and a wide variety of public services, including water supply and sewerage, for the population that can be expected to inhabit these newly formed or expanded towns and cities. Unfortunately, capital stock data on these various facilities are usually conveniently available only in highly aggregated form both by industry and geographical area. After investigating the data's availability, the author decided to focus attention on three highly aggregated sectors: state and local roads and highways, other state and local capital (total state and local capital less roads and highways), and owner-occupied housing. Farm capital used in production is not included in this study, but certainly merits close scrutiny in future research.

The Projection of Capital Stocks

After the capital stock data are obtained, the first task is to calculate and compare current capital stocks per square mile and per member of the rural population for the present Corn Belt and the area that would comprise the Corn Belt if climatic change took place. In addition, ratios of rural population to land area in both regions must be calculated and compared. If the data show that current capital requirements and man-land ratios are higher in the present Corn Belt than in the region defined as the Corn Belt with climatic change, (as indeed they do), it is clear that the shift of production from the old Corn Belt to the new implies some net capital formation and hence some net costs to society. Also, if the reduction of agricultural production in the most northerly regions causes the world market prices of foodstuffs to rise, capital requirements and their associated costs are likely to increase still further.

The capital stock projections for the old and new Corn Belts are made in several relatively simple steps. Several different increases (low, medium, and high) are hypothesized in the ratios of capital per unit of land area that are likely to occur in the absence of climatic change by the years 2000 and 2025. These ratios are then multiplied by the total area of each Corn Belt to obtain capital stocks in 2000 and 2025. In order to account for additional investments resulting from climatic change, capital stock in the new Corn Belt is estimated for the year 2025 using the projected capital per unit of land ratio of the old Corn Belt for 2025. It is assumed that shifts in capital stock attributable to climatic change begin in the year 2000 and continue until a steady state is reached in 2025.

Present Value Calculation

After the capital stocks for the new Corn Belt are estimated for 2000 and 2025, the next step is to calculate the present values of the net investments incurred over the period 1974 to 2025. The formula for obtaining present value is

$$PV = \sum_{i=1}^{n} \frac{I_i}{(1 + r)^i}$$

where I_i represents net investment during year i, r is the rate of discount, and n is the length of the time horizon. This calculation is done in two different ways. First, all annual net investments over the period 2000 to 2025 are discounted back to 1974 to obtain the present value in 1974 of the total net investment. Second, annual net investments over the period 2000 to 2025 are discounted back to yield the present value of the net investments in the year 2000. In each case, three separate calculations are made using interest rates of 3, 5, and 8%, respectively. Given this information, it is possible to calculate the costs of climatic change. They are estimated by subtracting the present value of net investments that would take place in the new Corn Belt without the occurrence of climatic change from the present value of net investments that would be required in the same area by 2025 capital per unit area ratios defined for the original Corn Belt under the assumption of no climatic change. It has been assumed that the land withdrawn from the growing of corn is shifted to dairy and wheat production. In addition, it has further been assumed that this shift does not reduce capital requirements in the old Corn Belt below the levels that were projected under the hypothesis of no climatic change. If this last assumption is not valid, the costs of climatic change may be overestimated because costs incurred in the new Corn Belt may be offset to some extent by reduced costs in the existing Corn Belt.

Annualization

In order to obtain some idea of the annual investment costs that climatic change would bring about, the net costs obtained above in the form of present values are annualized over the periods 1974 to 2025 and 2000 to 2025. A comparison of these annual net costs reveals the sensitivity of estimates to the long lag between the decision to build supersonic aircraft that might pollute the stratosphere and its eventual consequences. Specifically, it permits the annual costs as perceived by today's decision makers to be compared with the annual costs that are likely to be incurred by the population living during the period 2000 to 2025.

The Data and Their Limitations

State and Local Public Sector Capital Stocks

In order to project capital stocks into the future, some measure of current state and local capital stocks is required. Unfortunately, no such data are currently available. Therefore, it proved necessary to make rough estimates of gross capital stocks for both old and new Corn Belts for the end of 1970. These were constructed by cumulation by state of estimated annual data on state and local capital outlays for roads and highways and total annual state and local capital outlays *minus* capital outlays for roads and highways, obtained from the U.S. Department of Commerce, Bureau of the Census[1,2] for the years 1922 to 1970. Since prior to 1963, local data were published only as an aggregate for the entire nation, all figures for previous years are estimates based on multiplying state data by the national ratio of state plus local to state capital expenditures for the given type of capital during the given year. Thus, these figures are constructed under the assumption that the ratio of state to local government expenditures is constant among states during any given year. No data on capital stock retirements are available. However, it is reasonable to assume that they have been small compared to new construction, and therefore, would not significantly affect the estimated levels of capital stocks.

However, before the cumulation was carried out, the road and highway expenditures were deflated by the Bureau of Public Roads Highway Construction Cost Index, obtained from the U.S. Department of Commerce, Bureau of the Census,[3,4] adjusted to a 1971 base. In addition, the other capital outlay data (total capital outlays minus outlays for roads and highways) were deflated by the Engineering News-Record General Building Cost Index, again obtained from the U.S. Department of Commerce, Bureau of the Census,[3,4] adjusted to a 1971 base. After deflation, the data were cumulated to the end of 1970 to yield estimates of the two categories of gross capital stocks by state.

However, these figures cannot be aggregated for two entire Corn Belts since the borders of the latter are not coterminous with state boundaries. Therefore, four additional operations were carried out to obtain estimates of agriculturally related capital stocks for the entire Corn Belts. First, the statewide capital stocks were weighted by the ratios of the total 1970 populations of included Corn Belt counties lying within the state to the entire state population. This procedure, which is based on the assumption that per capita capital stocks are identical across all counties of the state, reduces the capital stock figures so they represent only capital stocks located in those portions of each state that lie within the Corn Belt. Second, based on the further assumption that per capita capital stocks are identical in both rural and urban areas, each of these estimates was weighted by the ratio of rural Corn Belt population within the state to total Corn Belt population within the state, yielding an estimate of rurally oriented capital stock. Third, in order to account for the capital stock that exists in urban areas that function as central places for the rural areas, each estimate was multiplied by 1.5, yielding a final estimate of rurally oriented Corn Belt capital stock by state. This ratio, which is an empirical estimate obtained from inspection of the data for primarily rural states, implies that for every two units of capital in the rural sector, one is necessary in the urban sector to service them. Fourth, all of the state estimates were then aggregated to obtain figures for both the existing Corn Belt and the new Corn Belt that results from climatic change.

It must be stressed that these capital stock estimates are only very rough approximations and are subject to many sources of error. However, it has been assumed that any errors that may exist tend to cancel each other out, rather than acting in a cumulatively reinforcing manner. A serious weakness of the concept state and local capital stock other than roads and highways must be pointed out. Although this figure includes schools, hospitals, and public utilities, it does not embrace all of the existing institutions. Obviously, private hospitals, schools, and utilities are not counted; yet, they may constitute a significant portion of the total stock in any given geographical area.

Housing

The housing data are superior in some respects to the public sector capital stock data, but inferior from other points of view. The basic data are the number and median value of owner-occupied units by county, taken from the 1970 Census of Housing published by the U.S. Department of Commerce, Bureau of the Census.[5] These two numbers were then multiplied together to obtain an estimate of the total value of owner-occupied housing in 1970 dollars by county. In addition, each estimate was then increased by 6% to convert it to 1971 dollars, as suggested by data published by the U.S. Department of Commerce, Bureau of the Census[4] showing that existing homes with FHA insured mortgages increased in median value by that amount during the period 1970 to 1971. Unfortunately, renter-occupied and vacant units could not be included in the analysis. Nevertheless, it is well known that in rural areas, towns, and small cities, the great majority of the total housing units falls into the owner-occupied category.

The data on market value of owner-occupied housing in 1971 dollars by county were then weighted by the ratio of 1970 county rural population to 1970 total county population. Then, each of these estimates was multiplied by 1.5 to account for housing stock in urban areas that function as central places for the rural areas. This ratio, which was also used in calculating public sector capital, is an empirical estimate obtained from inspection of the data for primarily rural counties. Next, the county data were aggregated to the state level. Finally, all of the state estimates were aggregated to obtain figures for both the existing Corn Belt and the new Corn Belt that results from climatic change.

Again, it must be emphasized that these capital stock estimates are only rough approximations and tend to be biased downward because the percentage of owner-occupied dwelling units tends to be higher in rural than in urban areas. Nevertheless, since the rural populations of the most populous highly urbanized counties tend to be quite small, and constitute only a very small proportion of the entire rural population for each state, the process of weighting aggregate values of owner-occupied housing by the ratio of rural to total population at the county level tends to minimize this bias.

Analysis

Area and Population

The area and rural population of both the existing Corn Belt and the Corn Belt with climatic change were aggregated from county data obtained from the U.S. Department of Commerce, Bureau of the Census[6] and are presented in Table 1.

Examination of the data reveals the area that constitutes the potential Corn Belt with climatic change currently has approximately 20% less land area, 24% less rural population, and a 5% smaller rural population density than the existing Corn Belt. In addition, it is well known that the soils in the southerly part of this region are generally less fertile than those in the northerly. Therefore, other things being equal, a shift in the Corn Belt would be almost certain to bring about a greater than 20% reduction in output of Corn Belt products.

Table 1.   Approximate Area, Rural Population, and Rural Population Density of the Existing Corn Belt and the Corn Belt with Climatic Change (1970)

	Area (square miles)	Rural Population (persons)	Rural Population Density (persons per square mile)
Existing Corn Belt	299,964	8,891,339	29.64
Corn Belt with climatic change	239,604	6,726,517	28.07

**Table 2.   Estimated Agriculturally Related Capital Stocks in the
Existing Corn Belt and the Corn Belt with Climatic Change on
January 1, 1971**

	Existing Corn Belt (billions of 1971 dollars)	Corn Belt with Climatic Change (billions of 1971 dollars)
State and local roads and highways	23.5	19.1
Other state and local	22.3	17.4
Owner-occupied housing	59.8	39.3

Estimated Capital Stocks

Following the procedures already discussed in this Chapter, estimates of
state and local road and highway, other state and local, and owner-occupied
housing capital stocks were made for both the present Corn Belt and the Corn
Belt with climatic change. They are presented in Table 2. Roads and highways
and other state and local capital are measured as gross stocks; housing is
measured in market values.

It appears that the estimated stock of owner-occupied housing in the existing
Corn Belt is about 50% larger than in the Corn Belt with climatic change.
However, it seems that the discrepancies between estimated stocks in the other
two sectors are much less. Nevertheless, the latter results appear to be suspect.
Certainly, one would expect that the levels of other state and local capital
should be roughly proportional to housing capital. In addition, the same thing
might be said of capital in roads and highways. Therefore, it appears that the
methods used to estimate Corn Belt capital from statewide data are not power-
ful enough to reflect intrastate differences between wealthy and poor counties.
On the other hand, the housing stocks, which were estimated from county data,
can be regarded as a good deal more reliable. Table 3 helps clarify these issues.

Upon examination of the housing data, one finds the ratios of housing stock
per unit of land and per member of the rural population are respectively about
20 and 15% greater in the existing Corn Belt. These results come about for two
reasons. First, the existing Corn Belt is a wealthier area, and therefore, has
better housing. Second, the capital stock figures include estimates of agriculturally related urban capital. In the preparation of the capital stock estimates, it became evident that agriculturally related urban populations in the
existing Corn Belt are very likely to be larger relative to rural populations than

Table 3. Ratios of Agriculturally Related Capital Stocks to Land Area and Rural Population

	Land Area		Rural Population	
	Existing Corn Belt (dollars per square mile)	Corn Belt with Climatic Change (dollars per square mile)	Existing Corn Belt (dollars per person)	Corn Belt with Climatic Change (dollars per person)
State and local	78,433	79,694	2646	2838
roads and highways	(85,696)	(70,508)	(2891)	(2511)
Other state	74,272	72,755	2505	2591
and local	(79,866)	(65,704)	(2694)	(2340)
Owner-occupied	199,194	164,155	6720	5847
housing				

in the area that constitutes the prospective future Corn Belt with climatic change. If this conjecture is indeed true, the existing Corn Belt must have larger aggregate capital stocks relative to both area and rural population.

The data for state and local roads and highways and other state and local capital that appear in Table 3 without parentheses do not show any consistent pattern. Given the suspicions expressed previously about the public sector capital stock estimates, it appears appropriate to adjust these data to make them consistent in relative magnitudes with the housing sector. This adjustment was carried out under two assumptions. First, it was decided to make the relative differences in the public sector capital stocks between the two Corn Belts equal to the same percentage difference manifested by the housing stock data. Second, the mean of the new estimates was specified equal to the mean of the existing estimates for each of the two public sectors. The new estimates are $25.706, $16.894, $23.957, and $15.743 billion for state and local roads and highways and other state and local capital in the existing and prospective Corn Belts, respectively. Ratios constructed from these two estimates appear in parentheses in Table 3. These data indicate that some fairly substantial direct costs will indeed be incurred by a shifting of the Corn Belt because the prospective Corn Belt with climatic change appears to be deficient in population and all three kinds of capital, both in the rural areas and agriculturally related urban areas.

Projections to 2000 and 2025

It is well known that the process of capital accumulation has been a significant force throughout the 20th century, continues unabated, and can be expected to persist into the future. Accumulation is usually accompanied by capital deepening in both the private and public sectors. Assuming that in the absence of climatic change, this accumulation and deepening will continue, the author has made three projections of Corn Belt capital stocks to 2000 and 2025. The low projection assumes increases in the capital per area ratio of 30% from 1971 to 2000 and another 20% to 2025, both calculated relative to the January 1, 1971 base. The medium projection assumes increases of 45 and 30%, and the high projection increases of 60 and 40%, respectively. These projections are presented in Table 4.

In order to calculate the direct costs of climatic change, it is also necessary to have estimates of capital stocks in the Corn Belt with climatic change, under the assumption that the climatic change has actually taken place. These estimates are made using capital per unit area ratios that can be calculated from the existing Corn Belt data in Table 4. These capital requirements are presented in Table 5.

First, it should be noted that the capital required in the new Corn Belt with climatic change in 2025 appears to be less than in the old Corn Belt without

Table 4. Projections of Agriculturally Oriented Corn Belt Capital Stocks in the Absence of Climatic Change

		Existing Corn Belt[a]			Corn Belt with Climatic Change[a]		
		1971	2000	2025	1971	2000	2025
Roads and	Low	25.7	33.4	38.5	16.9	22.0	25.4
highways	Medium	25.7	37.3	45.0	16.9	24.5	29.6
	High	25.7	41.1	51.4	16.9	27.0	33.8
Other state	Low	24.0	31.2	36.0	15.7	20.4	23.5
and local	Medium	24.0	34.8	42.0	15.7	22.8	27.5
	High	24.0	38.4	48.0	15.7	25.1	31.4
Housing	Low	59.8	77.7	89.7	39.3	51.1	59.0
	Medium	59.8	86.7	104.6	39.3	57.0	68.8
	High	59.8	95.7	119.6	39.3	62.9	78.6
Total	Low	109.5	142.3	164.2	71.9	93.5	107.9
	Medium	109.5	158.8	191.6	71.9	104.3	125.9
	High	109.5	175.2	219.0	71.9	115.0	143.8

[a] Billions of 1971 dollars.

Table 5. Capital Requirements for the Corn Belt with Climatic Change, Given a 5°F Fall in Temperature and Assuming That Adjustment to Climatic Changes Takes Place During the Period 2000 to 2025

		Stock in 1971	Stock in 2000[a]	Stock in 2025[a]	Change 2000 to 2025[a]
Roads and	Low	16.9	22.0	30.8	8.8
Highways	Medium	16.9	24.5	35.9	11.4
	High	16.9	27.0	41.1	14.1
Other State	Low	15.7	20.4	28.8	8.4
and Local	Medium	15.7	22.8	33.5	10.7
	High	15.7	25.1	38.3	13.2
Housing	Low	39.3	51.1	71.6	20.5
	Medium	39.3	57.0	83.6	26.6
	High	39.3	62.9	95.5	32.6
Total	Low	71.9	93.5	131.2	37.7
	Medium	71.9	104.3	153.0	48.7
	High	71.9	115.0	174.9	59.9

[a] Billions of 1971 dollars.

climatic change. This result suggests that the output of Corn Belt products will decrease during the period 2000 to 2025. A constant level of production could conceivably be maintained only at a cost of massive increases in capital used on the farm. However, estimation of these capital inputs would require a major research effort and cannot be attempted here. Given the information in Tables 4 and 5, it is a simple matter to subtract the net investments that would take place in the new Corn Belt without climatic change over the period 2000 to 2025 ($14.4, $21.6, and $28.8 billion under the low, medium, and high assumptions, respectively) from the corresponding totals in the last column of Table 5. The figures that result ($23.3, $27.1, and $31.1 billion) can be interpreted as the net investment costs attributable to climatic change under low, medium, and high assumptions, respectively, given a $2 \frac{7}{9}$°C fall in temperature. However, during the course of the research the decision was made to ascertain the consequences of a 1°C rather than a $2 \frac{7}{9}$°C fall in temperature. Since there was insufficient time for the agricultural specialists to correct their estimate of the shift in the Corn Belt, the above cost estimates were multiplied by 0.25 to yield figures based on a 1°C decrease. A factor of less than 0.36 (the reciprocal of 2 ⁷/₉) is assumed because investment requirements can be expected to increase more than linearly with distance in the southward direction as the fertility of the land decreases. Estimates of $5.8, $6.8, and $7.8 billion were obtained.

Present Value Calculation

Assuming net investment costs are incurred in 25 equal annual amounts over the period 2000 to 2025, these annual costs equal $0.233, $0.272, and $0.311 billions under the low, medium, and high assumptions, respectively. As mentioned earlier, present values of these net investment costs attributable to climatic change were calculated back to 1974 and 2000 in two separate operations for three different rates of interest. The results appear in Table 6.

It appears the present values in 1974 of the net investment costs attributable to climatic change lie in the range of $1.88 to $2.50 billion at a 3% rate of discount, $0.93 to $1.23 billion at a 5% rate of discount, and $0.33 to $0.45 billion at an 8% rate of discount. Thus, the costs appear to be significant, although very small relative to gross national product.

These costs may be overestimated if the capital requirements in that portion of the old Corn Belt which shifts to dairy and wheat production are reduced. However, there are two strong arguments for believing that they have actually been substantially underestimated. First, the costs of farm capital used in production have not been included in the analysis. Second, it appears that additional capital will be required to offset probable decreases in production. Since the new Corn Belt with a 1°C climatic change is somewhat smaller and is of lesser fertility than the old, assuming the capital stock levels projected in Table 5 implies a reduction in future corn production. In addition, if the production

Table 6.   Present Value of Net Investment in the New Corn Belt Attributable to a −1°C Climatic Change

		3% Discount Rate[a]	5% Discount Rate[a]	8% Discount Rate[a]
Present	Low	1.88	0.93	0.33
value	Medium	2.20	1.08	0.40
in 1974	High	2.50	1.23	0.45
Present	Low	4.08	3.28	2.48
value	Medium	4.73	3.83	2.90
in 2000	High	5.43	4.38	3.33

[a] Billions of 1971 dollars.

Table 7.   Annualized Present Values of Net Investment in the Corn Belt Attributable to a − 1°C Climatic Change

		3% Discount Rate[a]	5% Discount Rate[a]	8% Discount Rate[a]
1974	Low	0.072	0.051	0.027
to	Medium	0.085	0.059	0.033
2025	High	0.096	0.067	0.037
2000	Low	0.233	0.233	0.233
to	Medium	0.272	0.272	0.272
2025	High	0.311	0.311	0.311

[a] Billions of 1971 dollars.

of food in the more northern regions of the world is significantly reduced, world prices of foodstuffs are sure to rise rapidly. As a result of these forces working in the direction of reduced production and higher prices, it is almost certain that the labor and capital intensity of production in the Corn Belt would also increase markedly, and the costs attributable to climatic change would be significantly increased. Therefore, the figures presented in Table 6 should be considered as lower bounds of these costs.

Annualization Analysis

In order to obtain an idea of the annual costs brought about by climatic change, the present values of the net capital costs have been annualized over the periods 1974 to 2025 and 2000 to 2025. These estimates are given in Table 7.

Annualized costs vary from a possible low of $27 million spread over a 51-year period to a high of $311 million spread over a 25-year period. All the estimated costs associated with the shift in the Corn Belt are summarized in the table that appears in Appendix B.

Conclusion

Summary

This chapter has presented estimates of the net capital costs that might be incurred by a southward shift in the Corn Belt brought about by climatic changes. Low, medium, and high estimates of the direct costs were made. Present values of costs, discounted back to 1974, appear to vary from a low of $0.33 billion to a high of $2.5 billion.

Sensitivity

It is obvious that a number of heroic assumptions about the data have been made in this Chapter, and the potential for large errors in both measurement and forecasting is great. Nevertheless, it is encouraging that the estimated present values of costs show relatively little variability as projected increases in the capital per square mile ratio double. Finally, it appears that these cost estimates should be interpreted as lower bounds. Rising prices on world food markets attributable to reductions of production in the most northerly regions are likely to drive capital requirements and their associated costs up faster than projected. The construction of such revised estimates would require a knowledge of the reduction of agricultural production in northerly regions, the growth of world demand, the growth of exports, and the growth of domestic demand for food products. Finally, it would also require a forecast of shifts in the agricultural production function and patterns of factor utilization in agriculture.

Possible Further Research

Since the results reached in this essay indicate that the costs associated with a shift in the Corn Belt are not likely to be negligible, a more detailed study of the costs of possible shifts in agricultural production on a worldwide basis seems desirable. Such a study might be directed toward projecting the world demand and supply of agricultural products both with and without climatic change. Then, given what has already been learned about the likely geographical shifts in agricultural production, it would, in principle, be possible to estimate the costs associated with these shifts. In particular, changes and their accompanying costs in requirements for productive capital on the farm as well as in the public sector and housing could be investigated. Unfortunately, as has already been noted, the accurate capital stock data needed as inputs for a study of this magnitude are not readily available and could be obtained only at great expense.

Appendix A

Geographical Definition of the
Corn Belt before and after
Climatic Change (by States and Counties)

The Existing Corn Belt

Illinois—   Adams, Bond, Boone, Brown, Bureau, Calhoun, Carroll, Cass, Champaign, Christian, Clark, Clay, Clinton, Coles,

Cook, Crawford, Cumberland, De Kalb, De Witt, Douglas, Du Page, Edgar, Effingham, Fayette, Ford, Fulton, Greene, Grundy, Hancock, Henderson, Henry, Iroquois, Jasper, Jefferson, Jersey, Jo Daviess, Kane, Kankakee, Kendall, Knox, Lake, La Salle, Lawrence, Lee, Livingston, Logan, McDonough, McHenry, McLean, Macon, Macoupin, Madison, Marion, Marshall, Mason, Menard, Mercer, Monroe, Montgomery, Morgan, Moultrie, Ogle, Peoria, Perry, Piatt, Pike, Putnam, Randolph, Richland, Rock Island, St. Clair, Sangamon, Schuyler, Scott, Shelby, Stark, Stephenson, Tazewell, Vermillion, Warren, Washington, Wayne, Whiteside, Will, Winnebago, Woodford.

Indiana—   Adams, Allen, Batholomew, Benton, Blackford, Boone, Brown, Carroll, Cass, Clay, Clinton, Decatur, De Kalb, Delaware, Elkhart, Fayette, Fountain, Franklin, Fulton, Grant, Greene, Hamilton, Hancock, Hendricks, Henry, Howard, Huntington, Jasper, Jay, Johnson, Knox, Kosciusko, Lagrange, Lake, La Porte, Madison, Marion, Marshall, Miami, Monroe, Montgomery, Morgan, Newton, Noble, Owen, Parke, Porter, Pulaski, Putnam, Randolph, Rush, St. Joseph, Shelby, Starke, Steuben, Sullivan, Tippecanoe, Tipton, Union, Vermillion, Vigo, Wabash, Warren, Wayne, Wells, White, Whitley.

Iowa—   Adair, Adams, Allamakee, Appanoose, Audubon, Benton, Black Hawk, Boone, Bremer, Buchanan, Buena Vista, Butler, Calhoun, Carroll, Cass, Cedar, Cerro Gordo, Cherokee, Chickasaw, Clarke, Clay, Clayton, Clinton, Crawford, Dallas, Davis, Decatur, Delaware, Des Moines, Dickinson, Dubuque, Emmet, Fayette, Floyd, Franklin, Fremont, Greene, Grundy, Guthrie, Hamilton, Hancock, Hardin, Harrison, Henry, Howard, Humboldt, Ida, Iowa, Jackson, Jasper, Jefferson, Johnson, Jones, Keokuk, Kossuth, Lee, Linn, Louisa, Lucas, Lyon, Madison, Mahaska, Marion, Marshall, Mills, Mitchell, Monona, Monroe, Montgomery, Muscatine, O'Brien, Osceola, Page, Palo Alto, Plymouth, Pocahontas, Polk, Pottawattamie, Poweshiek, Ringgold, Sac, Scott, Shelby, Sioux, Story, Tama, Taylor, Union, Van Buren, Wapello, Warren, Washington, Wayne, Webster, Winnebago, Winneshiek, Woodbury, Worth, Wright.

Kansas—   Allen, Anderson, Atchison, Bourbon, Brown, Coffey, Doniphan, Douglas, Franklin, Jackson, Jefferson, Johnson,

Leavenworth, Linn, Marshall, Miami, Nemaha, Osage, Pottawatomie, Shawnee, Wabaunsee, Wyandotte.

Michigan—   Allegan, Barry, Berrien, Branch, Calhoun, Cass, Eaton, Hillsdale, Jackson, Kalamazoo, Lenawee, Monroe, St. Joseph, Van Buren, Washtenaw.

Minnesota—   Big Stone, Blue Earth, Brown, Carver, Chippewa, Cottonwood, Dakota, Dodge, Douglas, Faribault, Fillmore, Freeborn, Goodhue, Grant, Hennepin, Houston, Jackson, Kandiyohi, Lac qui Parle, Le Sueur, Lincoln, Lyon, McLeod, Martin, Meeker, Mower, Murray, Nicollet, Nobles, Olmsted, Pipestone, Pope, Redwood, Renville, Rice, Rock, Scott, Sibley, Stearns, Steele, Stevens, Swift, Traverse, Wabasha, Waseca, Watowan, Wilkin, Winona, Wright, Yellow Medicine.

Missouri—   Adair, Andrew, Atchison, Audrain, Bates, Benton, Boone, Buchanan, Caldwell, Callaway, Camden, Carroll, Cass, Cedar, Chariton, Clark, Clay, Clinton, Cole, Cooper, Crawford, Dallas, Daviess, De Kalb, Franklin, Gasconade, Gentry, Grundy, Harrison, Henry, Hickory, Holt, Howard, Jackson, Jefferson, Johnson, Knox, Laciede, Lafayette, Lewis, Lincoln, Linn, Livingston, Macon, Maries, Marion, Mercer, Miller, Moniteau, Monroe, Montgomery, Morgan, Nodaway, Osage, Pettis, Phelps, Pike, Platte, Polk, Pulaski, Putnam, Ralis, Randolph, Ray, St. Charles, St. Clair, St. Louis, St. Louis City, Schuyler, Scotland, Shelby, Sullivan, Vernon, Warren, Washington, Worth.

Nebraska—   Burt, Butler, Cass, Cedar, Colfax, Cuming, Dakota, Dixon, Dodge, Douglas, Fillmore, Gage, Jefferson, Johnson, Knox, Lancaster, Madison, Nemaha, Otoe, Pawnee, Pierce, Platte, Polk, Richardson, Saline, Sarpy, Saunders, Seward, Stanton, Thayer, Thurston, Washington, Wayne, York.

North Dakota—Dickey, Ransom, Richland, Sargent.

Ohio—   Allen, Auglaize, Butler, Champaign, Clark, Crawford, Darke, Defiance, Delaware, Fulton, Greene, Hancock, Hardin, Henry, Logan, Madison, Marion, Mercer, Miami, Montgomery, Ottawa, Paulding, Preble, Putnam, Sandusky, Seneca, Shelby, Union, Van Wert, Warren, Williams, Wood, Wyandot.

South Dakota—Beadle, Bon Homme, Brookings, Brown, Clark, Clay, Codington, Davison, Day, Deuel, Grant, Hamlin, Hanson, Hutchinson, Kingsbury, Lake, Lincoln, McCook, Marshall,

Miner, Minnehaha, Moody, Roberts, Sanborn, Spink, Turner, Union, Yankton.

Wisconsin— Crawford, Dane, Grant, Green, Iowa, Jefferson, Kenosha, La Crosse, Lafayette, Milwaukee, Racine, Richland, Rock, Sauk, Vernon, Walworth, Waukesha.

Corn Belt with Climatic Change

Arkansas— Baxter, Benton, Boone, Carroll, Clay, Cleburne, Craighead, Crawford, Franklin, Fulton, Greene, Independence, Izard, Jackson, Johnson, Lawrence, Madison, Marion, Mississippi, Newton, Polk, Randolph, Searcy, Sharp, Stone, Van Buren, Washington.

Illinois— Adams, Alexander, Bond, Brown, Calhoun, Cass, Champaign, Christian, Clark, Clay, Clinton, Coles, Crawford, Cumberland, De Witt, Douglas, Edgar, Edwards, Effingham, Fayette, Franklin, Fulton, Gallatin, Greene, Hamilton, Hancock, Hardin, Henderson, Jackson, Jasper, Jefferson, Jersey, Johnson, Lawrence, Logan, McDonough, McLean, Macon, Macoupin, Madison, Marion, Mason, Massac, Menard, Monroe, Montgomery, Morgan, Moultrie, Perry, Piatt, Pike, Pope, Pulaski, Randolph, Richland, St. Clair, Saline, Sangamon, Schuyler, Scott, Shelby, Tazewell, Union, Vermilion, Wabash, Warren, Washington, Wayne, White, Williamson.

Indiana— Bartholomew, Boone, Brown, Clark, Clay, Clinton, Crawford, Daviess, Dearborn, Decatur, Delaware, Dubois, Fayette, Floyd, Fountain, Franklin, Gibson, Greene, Hamilton, Hancock, Harrison, Hendricks, Henry, Jackson, Jefferson, Jennings, Johnson, Knox, Lawrence, Madison, Marion, Martin, Monroe, Montgomery, Morgan, Ohio, Orange, Owen, Parke, Perry, Pike, Posey, Putnam, Randolph, Ripley, Rush, Scott, Shelby, Spencer, Sullivan, Switzerland, Tippecanoe, Tipton, Union, Vanderburgh, Vermillion, Vigo, Warren, Warrick, Washington, Wayne.

Iowa— Adair, Adams, Appanoose, Cass, Clarke, Davis, Decatur, Des Moines, Fremont, Harrison, Henry, Jefferson, Keokuk, Lee, Lucas, Madison, Mahaska, Marion, Mills, Monroe, Montgomery, Page, Pottawattamie, Ringgold, Shelby, Taylor, Union, Van Buren, Wapello, Warren, Wayne.

Kansas— Allen, Anderson, Atchison, Bourbon, Brown, Cherokee, Coffey, Crawford, Doniphan, Douglas, Elk, Franklin, Greenwood, Jackson, Jefferson, Johnson, Labette, Leavenworth, Linn, Lyon, Marshall, Miami, Montgomery, Nemaha, Neosho, Osage, Pottawatomie, Shawnee, Wabaunsee, Wilson, Woodson, Wyandotte.

Kentucky— Adair, Allen, Anderson, Ballard, Barren, Boone, Bourbon, Boyle, Bracken, Breckinridge, Bullitt, Butler, Caldwell, Calloway, Campbell, Carlisle, Carroll, Casey, Christian, Clark, Crittenden, Daviess, Edmonson, Fayette, Franklin, Fulton, Gallatin, Garrard, Grant, Graves, Grayson, Green, Greenup, Hancock, Hardin, Harrison, Hart, Henderson, Henry, Hickman, Hopkins, Jefferson, Jessamine, Kenton, Larue, Lincoln, Livingston, Logan, Lyon, McCracken, McLean, Madison, Marion, Marshall, Mason, Meade, Mercer, Metcalfe, Monroe, Muhlenberg, Nelson, Nicholas, Ohio, Oldham, Owen, Pendleton, Robertson, Scott, Shelby, Simpson, Spencer, Taylor, Todd, Trigg, Trimble, Union, Warren, Washington, Webster, Woodford.

Missouri— Adair, Andrew, Atchison, Audrain, Barry, Barton, Bates, Benton, Bollinger, Boone, Buchanan, Butler, Caldwell, Callaway, Camden, Cape Girardeau, Carroll, Carter, Cass, Cedar, Chariton, Christian, Clark, Clay, Clinton, Cole, Cooper, Crawford, Dade, Dallas, Daviess, De Kalb, Dent, Douglas, Dunklin, Franklin, Gasconade, Gentry, Greene, Grundy, Harrison, Henry, Hickory, Holt, Howard, Howell, Iron, Jackson, Jasper, Jefferson, Johnson, Knox, Laciede, Lafayette, Lawrence, Lewis, Lincoln, Linn, Livingston, McDonald, Macon, Madison, Maries, Marion, Mercer, Miller, Mississippi, Moniteau, Monroe, Montgomery, Morgan, New Madrid, Newton, Nodaway, Oregon, Osage, Ozark, Pemiscott, Perry, Pettis, Phelps, Pike, Platte, Polk, Pulaski, Putnam, Ralis, Randolph, Ray, Reynolds, Ripley, St. Charles, St. Clair, St. Francois, St. Louis, St. Louis City, Ste. Genevieve, Saline, Schuyler, Scotland, Scott, Shannon, Shelby, Stoddard, Stone, Sullivan, Taney, Texas, Vernon, Warren, Washington, Wayne, Webster, Worth, Wright.

Nebraska— Butler, Cass, Dodge, Douglas, Fillmore, Gage, Jefferson, Johnson, Lancaster, Nemaha, Otoe, Pawnee, Richardson, Saline, Sarpy, Saunders, Seward, Thayer, Washington.

Ohio— Adams, Brown, Butler, Clark, Clermont, Clinton, Darke, Fayette, Greene, Hamilton, Highland, Miami, Montgomery, Preble, Warren.

Oklahoma—   Adair, Cherokee, Craig, Delaware, Mayes, Nowata, Ottawa, Rogers, Wagoner, Washington.

Tennessee—   Benton, Carroll, Cheatham, Crockett, Dickson, Dyer, Gibson, Henry, Houston, Humphreys, Lake, Lauderdale, Montgomery, Obion, Robertson, Stewart, Sumner, Weakley.

Appendix B

Summary Table

Table B.1. Present Values and Annualized Costs, Medium Projection (Billions of 1971 dollars)

Discount Rate	3%	5%	8%
1974 Present Value of Costs	2.200	1.080	0.400
Annualized Costs 1974 to 2025	0.085	0.059	0.033
2000 Present Value of Costs	4.730	3.830	2.900
Annualized Costs 2000 to 2025	0.272	0.272	0.272

References

1. U.S. Department of Commerce, Bureau of the Census, *Historical Statistics on State and Local Governmental Finances, 1902–1953,* Washington, D.C.: U.S. Government Printing Office, 1955.

2. U. S. Department of Commerce, Bureau of the Census, *Summary of Governmental Finances, and Governmental Finances,* Vols. G-GF54 to G-GF71, Washington, D.C.: U.S. Government Printing Office, 1955–1972.

3. U.S. Department of Commerce, Bureau of the Census, *Historical Statistics of the United States, Colonial Times to 1957,* and *Continuation to 1962 and Revisions,* Washington, D.C.: U.S. Government Printing Office, 1960 and 1965.

4. U.S. Department of Commerce, Bureau of the Census, *Statistical Abstract of the United States,* Vols. 84–93, Washington, D.C.: U.S. Government Printing Office, 1963–1972.

5. U.S. Department of Commerce, Bureau of the Census, *Census of Housing,* 1970, Vol. 1, *Housing Characteristics for States, Cities, and Counties,* Parts 5, 15–19, 24, 25, 27, 29, 36–38, 43, 44, and 51. Washington, D.C.: U.S. Government Printing Office, 1972.

6. U.S. Department of Commerce, Bureau of the Census, *County and City Data Book, 1972: A Statistical Abstract Supplement,* Washington, D.C.: U.S. Government Printing Office, 1973.

INDEX